This report contains the collective views of an international group of experts and does not necessarily represent the decisions or the stated policy of the United Nations Environment Programme, the International Labour Organisation, or the World Health Organization.

**Environmental Health Criteria 185**

# CHLORENDIC ACID AND ANHYDRIDE

First draft prepared by Dr G.J. van Esch,
Bilthoven, Netherlands

Published under the joint sponsorship of
the United Nations Environment Programme,
the International Labour Organisation,
and the World Health Organization

World Health Organization
Geneva, 1996

The **International Programme on Chemical Safety (IPCS)** is a joint venture of the United Nations Environment Programme, the International Labour Organisation, and the World Health Organization. The main objective of the IPCS is to carry out and disseminate evaluations of the effects of chemicals on human health and the quality of the environment. Supporting activities include the development of epidemiological, experimental laboratory, and risk-assessment methods that could produce internationally comparable results, and the development of manpower in the field of toxicology. Other activities carried out by the IPCS include the development of know-how for coping with chemical accidents, coordination of laboratory testing and epidemiological studies, and promotion of research on the mechanisms of the biological action of chemicals.

WHO Library Cataloguing in Publication Data

Chlorendic acid and anhydride.

(Environmental health criteria ; 185)

1.Insecticides, Organochlorine 2.Environmental exposure
I.Series

ISBN 92 4 157185 3 (NLM Classification: WA 240)
ISSN 0250-863X



PRINTED IN FINLAND
96/10846 – VAMMALA – 5500

## CONTENTS

ENVIRONMENTAL HEALTH CRITERIA FOR CHLORENDIC ACID AND ANHYDRIDE

## NOTE TO READERS OF THE CRITERIA MONOGRAPHS

Every effort has been made to present information in the criteria monographs as accurately as possible without unduly delaying their publication. In the interest of all users of the Environmental Health Criteria monographs, readers are requested to communicate any errors that may have occurred to the Director of the International Programme on Chemical Safety, World Health Organization, Geneva, Switzerland, in order that they may be included in corrigenda.

* * *

A detailed data profile and a legal file can be obtained from the International Register of Potentially Toxic Chemicals, Case postale 356, 1219 Châtelaine, Geneva, Switzerland (Telephone No. 9799111).

* * *

This publication was made possible by grant number 5 U01 ES02617-15 from the National Institute of Environmental Health Sciences, National Institutes of Health, USA, and by financial support from the European Commission.

Environmental Health Criteria

PREAMBLE

## Objectives

In 1973 the WHO Environmental Health Criteria Programme was initiated with the following objectives:

(i) to assess information on the relationship between exposure to environmental pollutants and human health, and to provide guidelines for setting exposure limits;

(ii) to identify new or potential pollutants;

(iii) to identify gaps in knowledge concerning the health effects of pollutants;

(iv) to promote the harmonization of toxicological and epidemiological methods in order to have internationally comparable results.

The first Environmental Health Criteria (EHC) monograph, on mercury, was published in 1976 and since that time an ever-increasing number of assessments of chemicals and of physical effects have been produced. In addition, many EHC monographs have been devoted to evaluating toxicological methodology, e.g., for genetic, neurotoxic, teratogenic and nephrotoxic effects. Other publications have been concerned with epidemiological guidelines, evaluation of short-term tests for carcinogens, biomarkers, effects on the elderly and so forth.

Since its inauguration the EHC Programme has widened its scope, and the importance of environmental effects, in addition to health effects, has been increasingly emphasized in the total evaluation of chemicals.

The original impetus for the Programme came from World Health Assembly resolutions and the recommendations of the 1972 UN Conference on the Human Environment. Subsequently the work became an integral part of the International Programme on Chemical Safety (IPCS), a cooperative programme of UNEP, ILO and WHO. In this manner, with the strong support of the new 14 partners, the importance of occupational health and environmental effects was fully recognized. The EHC monographs have become widely established, used and recognized throughout the world.

The recommendations of the 1992 UN Conference on Environment and Development and the subsequent establishment of the Intergovernmental Forum on Chemical Safety with the priorities for action in the six programme areas of Chapter 19, Agenda 21, all lend further weight to the need for EHC assessments of the risks of chemicals.

## Scope

The criteria monographs are intended to provide critical reviews on the effect on human health and the environment of chemicals and of combinations of chemicals and physical and biological agents. As such, they include and review studies that are of direct relevance for the evaluation. However, they do not describe *every* study carried out. Worldwide data are used and are quoted from original studies, not from abstracts or reviews. Both published and unpublished reports are considered and it is incumbent on the authors to assess all the articles cited in the references. Preference is always given to published data. Unpublished data are only used when relevant published data are absent or when they are pivotal to the risk assessment. A detailed policy statement is available that describes the procedures used for unpublished proprietary data so that this information can be used in the evaluation without compromising its confidential nature (WHO (1990) Revised Guidelines for the Preparation of Environmental Health Criteria Monographs. PCS/90.69, Geneva, World Health Organization).

In the evaluation of human health risks, sound human data, whenever available, are preferred to animal data. Animal and *in vitro* studies provide support and are used mainly to supply evidence missing from human studies. It is mandatory that research on human subjects is conducted in full accord with ethical principles, including the provisions of the Helsinki Declaration.

The EHC monographs are intended to assist national and international authorities in making risk assessments and subsequent risk management decisions. They represent a thorough evaluation of risks and are not, in any sense, recommendations for regulation or standard setting. These latter are the exclusive purview of national and regional governments.

## Content

The layout of EHC monographs for chemicals is outlined below.

- Summary - a review of the salient facts and the risk evaluation of the chemical
- Identity - physical and chemical properties, analytical methods
- Sources of exposure
- Environmental transport, distribution and transformation
- Environmental levels and human exposure
- Kinetics and metabolism in laboratory animals and humans
- Effects on laboratory mammals and *in vitro* test systems
- Effects on humans
- Effects on other organisms in the laboratory and field
- Evaluation of human health risks and effects on the environment
- Conclusions and recommendations for protection of human health and the environment
- Further research
- Previous evaluations by international bodies, e.g., IARC, JECFA, JMPR

## Selection of chemicals

Since the inception of the EHC Programme, the IPCS has organized meetings of scientists to establish lists of priority chemicals for subsequent evaluation. Such meetings have been held in: Ispra, Italy, 1980; Oxford, United Kingdom, 1984; Berlin, Germany, 1987; and North Carolina, USA, 1995. The selection of chemicals has been based on the following criteria: the existence of scientific evidence that the substance presents a hazard to human health and/or the environment; the possible use, persistence, accumulation or degradation of the substance shows that there may be significant human or environmental exposure; the size and nature of populations at risk (both human and other species) and risks for environment; international concern, i.e. the substance is of major interest to several countries; adequate data on the hazards are available.

If an EHC monograph is proposed for a chemical not on the priority list, the IPCS Secretariat consults with the Cooperating Organizations and all the Participating Institutions before embarking on the preparation of the monograph.

## Procedures

The order of procedures that result in the publication of an EHC monograph is shown in the flow chart. A designated staff member of IPCS, responsible for the scientific quality of the document, serves as Responsible Officer (RO). The IPCS Editor is responsible for layout and language. The first draft, prepared by consultants or, more usually, staff from an IPCS Participating Institution, is based initially on data provided from the International Register of Potentially Toxic Chemicals, and reference data bases such as Medline and Toxline.

The draft document, when received by the RO, may require an initial review by a small panel of experts to determine its scientific quality and objectivity. Once the RO finds the document acceptable as a first draft, it is distributed, in its unedited form, to well over 150 EHC contact points throughout the world who are asked to comment on its completeness and accuracy and, where necessary, provide additional material. The contact points, usually designated by governments, may be Participating Institutions, IPCS Focal Points, or individual scientists known for their particular expertise. Generally some four months are allowed before the comments are considered by the RO and author(s). A second draft incorporating comments received and approved by the Director, IPCS, is then distributed to Task Group members, who carry out the peer review, at least six weeks before their meeting.

The Task Group members serve as individual scientists, not as representatives of any organization, government or industry. Their function is to evaluate the accuracy, significance and relevance of the information in the document and to assess the health and environmental risks from exposure to the chemical. A summary and recommendations for further research and improved safety aspects are also required. The composition of the Task

EHC PREPARATION FLOW CHART

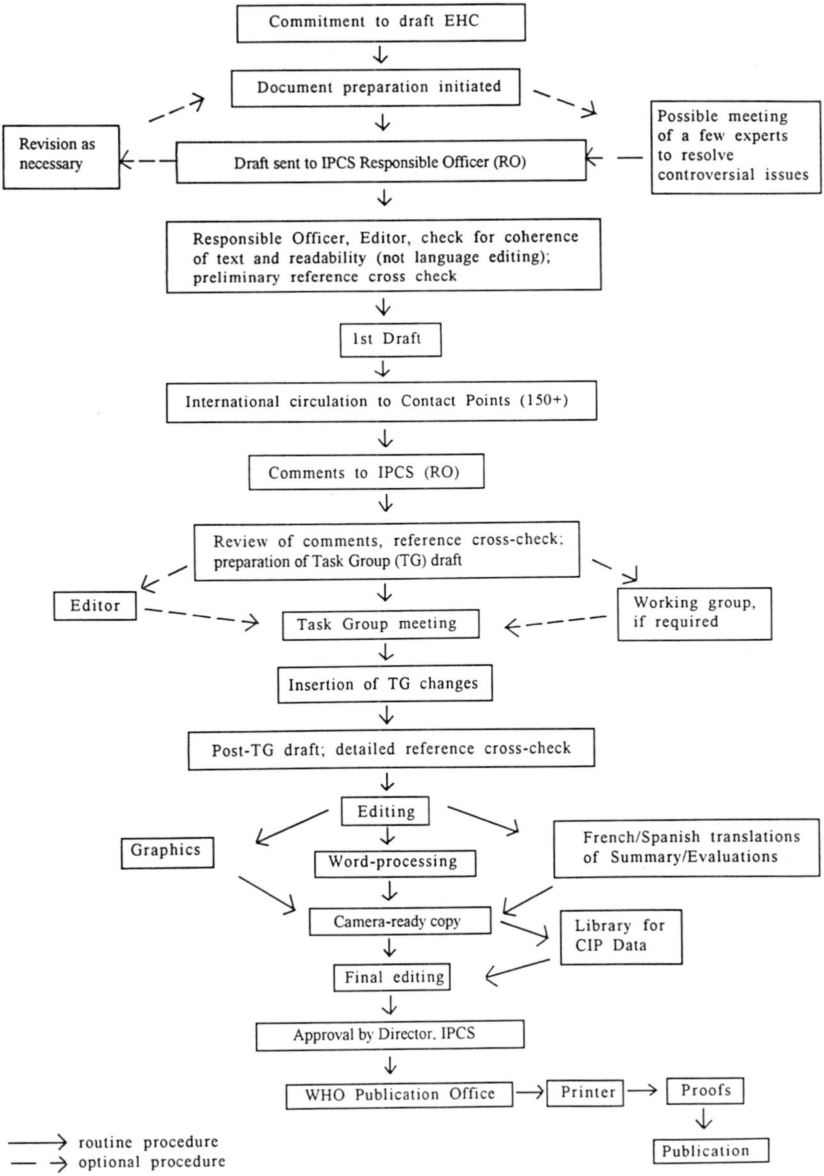

Group is dictated by the range of expertise required for the subject of the meeting and by the need for a balanced geographical distribution.

The three cooperating organizations of the IPCS recognize the important role played by nongovernmental organizations. Representatives from relevant national and international associations may be invited to join the Task Group as observers. While observers may provide a valuable contribution to the process, they can only speak at the invitation of the Chairperson. Observers do not participate in the final evaluation of the chemical; this is the sole responsibility of the Task Group members. When the Task Group considers it to be appropriate, it may meet *in camera*.

All individuals who as authors, consultants or advisers participate in the preparation of the EHC monograph must, in addition to serving in their personal capacity as scientists, inform the RO if at any time a conflict of interest, whether actual or potential, could be perceived in their work. They are required to sign a conflict of interest statement. Such a procedure ensures the transparency and probity of the process.

When the Task Group has completed its review and the RO is satisfied as to the scientific correctness and completeness of the document, it then goes for language editing, reference checking, and preparation of camera-ready copy. After approval by the Director, IPCS, the monograph is submitted to the WHO Office of Publications for printing. At this time a copy of the final draft is sent to the Chairperson and Rapporteur of the Task Group to check for any errors.

It is accepted that the following criteria should initiate the updating of an EHC monograph: new data are available that would substantially change the evaluation; there is public concern for health or environmental effects of the agent because of greater exposure; an appreciable time period has elapsed since the last evaluation.

All Participating Institutions are informed, through the EHC progress report, of the authors and institutions proposed for the drafting of the documents. A comprehensive file of all comments received on drafts of each EHC monograph is maintained and is available on request. The Chairpersons of Task Groups are briefed before each meeting on their role and responsibility in ensuring that these rules are followed.

## IPCS TASK GROUP ON ENVIRONMENTAL HEALTH CRITERIA FOR CHLORENDIC ACID AND ANHYDRIDE

*Members*

Dr L.A. Albert, Xalapa, Veracruz, Mexico (*Vice-Chairman*)

Dr S.K. Kashyap, Director, National Institute of Occupational Health, Ahmedabad, India

Mr H. Malcolm, Institute of Terrestrial Ecology, Monks Wood Experimental Station, Huntingdon, United Kingdom (part-time)

Dr K. Peltonen, Institute of Occupational Health, Helsinki, Finland

Professor Wai-On Phoon, Worksafe Australia and Department of Occupational Health, University of Sydney, Sydney, Australia (*Chairman*)

Mr D.J. Reisman, US Environmental Protection Agency, Cincinnati, USA

Dr E. Søderlund, National Institute of Public Health, Oslo, Norway

Dr G.J. van Esch, Bilthoven, Netherlands (*Rapporteur*)

*Observers*

Mr P.C. Schreiber, Occidental Chemical, Genk, Belgium

*Secretariat*

Dr K.W. Jager, International Programme on Chemical Safety, World Health Organization, Geneva, Switzerland

Mr J. Wilbourn, International Agency for Research on Cancer (IARC), Lyon, France

## ENVIRONMENTAL HEALTH CRITERIA FOR CHLORENDIC ACID AND ANHYDRIDE

A WHO Task Group on Environmental Health Criteria for Chlorendic Acid and Anhydride met at the World Health Organization, Geneva, from 12 to 16 December 1994. Dr K.W. Jager, IPCS, welcomed the participants on behalf of Dr M. Mercier, Director of the IPCS, and the three IPCS cooperating organizations (UNEP/ILO/WHO). The Group reviewed and revised the draft monograph and made an evaluation of the risks for human health and the environment from exposure to chlorendic acid and anhydride.

The first draft of the monograph was prepared by Dr G.J. van Esch, Bilthoven, Netherlands. The second draft, incorporating comments received following circulation of the first draft to the IPCS contact points for Environmental Health Criteria monographs, was prepared by the IPCS Secretariat.

Dr K.W. Jager and Dr P.G. Jenkins, both of the IPCS Central Unit, were responsible for the scientific content of the monograph and technical editing, respectively.

The fact that industry made available to the IPCS and the Task Group their proprietary toxicological information on the products under discussion is gratefully acknowledged. This allowed the Task Group to make its evaluation on a more complete database.

The effort of all who helped in the preparation and the finalization of the document is gratefully acknowledged.

# INTRODUCTION

Chlorendic acid and anhydride are important flame retardants. However, it should be noted that the IPCS is preparing several other EHC monographs on flame retardants, which will provide additional information relevant to chlorendic acid and anhydride.

There will be one monograph, *Flame Retardants - A General Introduction* (in preparation), giving a general introduction to the use, the mode of action, and the potential risks of flame retardants. It will list the substances used as flame retardants and give a general indication of the data available on them.

Some flame retardants in wide use have been discussed in separate monographs, e.g., EHC 162: *Brominated diphenyl ethers* (IPCS, 1994a) and EHC 172: *Tetrabromobisphenol-A and derivatives* (IPCS, 1995a).

Certain flame retardants considered hazardous for humans and the environment have also been reviewed in separate monographs e.g., EHC 152: *Polybrominated biphenyls* (IPCS, 1994b), and EHC 173: *Tris- and bis(2,3-dibromopropyl) phosphate* (IPCS, 1995b).

Because of the possibility of the formation of halogenated dibenzodioxins and dibenzofurans under certain circumstances, such as pyrolysis, the following monographs have been developed: EHC 88: *Polychlorinated dibenzodioxins and dibenzofurans* (IPCS, 1989) and *Polybrominated dibenzodioxins and dibenzofurans* (in preparation).

The reader should consult these monographs for further information.

Whatever their use, flame retardants will ultimately end up in the environment, either as such or as breakdown products. The ultimate breakdown products and their levels will differ according to whether they have been used as reactives or as additive flame retardants.

In order to make a proper hazard assessment for humans and the environment, it is essential that, apart from toxicity and ecotoxicity data, the following are available:

- data on the ultimate fate of the substance under various use and disposal conditions, including incineration, and on its breakdown products; and

- adequate data on the persistence and bioaccumulation/biomagnification of the substance and its breakdown products.

# 1. SUMMARY AND EVALUATION; CONCLUSIONS AND RECOMMENDATIONS

## 1.1 Summary and evaluation

### *1.1.1 Physical and chemical properties*

Chlorendic acid (commercial grade 99.5%) and chlorendic anhydride (technical grade 97%) are closely related white crystalline materials. Structurally they are closely related to the chlorinated cyclodiene insecticides. When chlorendic acid is heated in an open system it loses water, forming chlorendic anhydride. Chlorendic anhydride can be hydrolysed rapidly to chlorendic acid. The melting points range from 208 °C (acid) to 235 °C (anhydride).

### *1.1.2 Production and use*

Chlorendic acid and chlorendic anhydride are mostly used as reactive flame retardants in polyester resins and plasticizers for electrical systems and paints, and in fibreglass-reinforced resins for chemical process equipment. In the textile industry they have been used in the past as finishing treatment for wool fabrics and carpets.

The combined worldwide production for chlorendic acid and anhydride is at present around 4000 tonnes per year.

### *1.1.3 Environmental transport, distribution and transformation*

Chlorendic acid may be released via hydrolytic degradation of polyesters and as an oxidation product of chlorinated cyclodiene insecticides.

Ultraviolet light degrades chlorendic acid with a half-life of 16 days in a solid thin layer and 5 days in an aqueous solution. In soil the half-life ranges from 140 to 280 days. Chlorendic acid is fairly persistent in soil, although the data available on this subject are inadequate.

No data are available on bioaccumulation and biomagnification potential or on ultimate fate following use.

Exposure to chlorendic anhydride is likely to lead to exposure to chlorendic acid due to hydrolysis of the former.

### 1.1.4 Environmental levels and human exposure

Chlorendic acid has been found in landfill leachate at concentrations up to 455 mg/litre.

### 1.1.5 Kinetics and metabolism in laboratory animals

After oral and intravenous administration of radiolabelled chlorendic acid to rats, the substance was rapidly distributed throughout the body and rapidly metabolized. More than 90% of the radiolabel was excreted within 24 h in the faeces, mainly in a conjugated form. Only 3-6% was excreted in urine. The highest concentrations of radiolabel were found in adipose tissue, liver, kidneys, whole blood and lung.

A similar pattern was found in the rat in a gavage study with chlorendic anhydride. The half-life of the radiolabel in this latter study was less than 2 days, except for fat, where it was 22.5 days.

No data on kinetics following the dermal or inhalation routes are available.

### 1.1.6 Effects on laboratory mammals and *in vitro* test systems

The acute oral toxicity of chlorendic acid is low; the $LD_{50}$ for the rat is 1770 mg/kg body weight. In the case of chlorendic anhydride, the oral $LD_{50}$ for rats is 2480 mg/kg body weight.

The acute dermal $LD_{50}$ of crude chlorendic anhydride in rabbits is > 3000 mg/kg body weight.

The 4-h dust inhalation $LC_{50}$ of chlorendic acid in rats is > 0.79 mg/litre.

Chlorendic acid and anhydride are skin irritants and severe eye and respiratory tract irritants in the rabbit. Chlorendic anhydride is a skin sensitizer in guinea-pigs, but one test with chlorendic acid gave a negative result.

A no-observed-effect level (NOEL) of 2500 mg/kg diet (equivalent to 250 mg/kg body weight) was found in a 13-week feeding study using chlorendic acid in mice; in rats the NOEL in a similar feeding study was 1250 mg/kg diet (equivalent to 62.5 mg/kg body weight). At higher doses growth depression was significant and microscopic changes were seen in the liver.

In a 90-day feeding study on rats with chlorendic anhydride, the NOEL was 125 mg/kg body weight per day, and in a 3-week dermal study it was 100 mg/kg body weight/day (apart from skin irritation). No NOEL could be established in a 28-day dust inhalation study on rats.

A teratogenicity study on rats with chlorendic anhydride, administered orally by gavage at dose levels of up to 400 mg/kg body weight on gestational days 6-15, showed maternal toxicity but no teratogenic effects.

Chlorendic acid was tested for mutagenic potential in five strains of *Salmonella typhimurium* in the presence and absence of an exogenous metabolism system. Negative results were obtained at dose levels up to 7690 μg/plate. A mouse lymphoma mutation assay in the absence of an exogenous metabolism system was positive. The dose levels tested were 1300 to 1700 μg/ml. The highest dose level was cytotoxic.

Chlorendic acid was positive in a transformation assay using BALB/c3T3 cells without metabolic activation, and was negative in a test for sex-linked recessive lethal mutations in male *Drosophila melanogaster*. Chlorendic acid did not give an increase of replicative DNA synthesis after oral or subcutaneous application of 450 or 900 mg/kg body weight to F-344 rats.

In tests of chlorendic anhydride with five strains of *Salmonella typhimurium* and one strain of the yeast *Saccharomyces cerevisiae*, at dose levels of up to 7500 μg/plate, no mutagenic potential was found. It did not induce forward mutation in a mouse lymphoma test and was negative in a transformation assay in BALB/c3T3 cells. It produced significant unscheduled DNA synthesis in human WI-38 cells. In a dominant lethal assay, mice were exposed to single doses of up to 223 mg chlorendic anhydride/kg body weight, followed by a breeding period of 7 weeks. Only a statistically significant decrease in the fertility index, relative to controls, for all females mated to treated males during week 5, and for females mated to mid-dose level males during week 4, was observed. No effects were seen on the number of implantations, resorptions or dead implants. It was concluded that this test yielded negative results, although the study design was inadequate.

Chlorendic acid was tested for carcinogenic potential in F-344/N rats at dose levels of 620 and 1250 mg/kg diet (equivalent to 31 and 62.5 mg/kg body weight). In addition to significant non-neoplastic changes in a number of organs, such as cystic degeneration and focal cellular changes, and bile duct hyperplasia in the liver, increases in the incidence of hepatocellular adenomas in treated males and hepatocellular adenomas and carcinomas, significant at the highest dose level, in females were found. Furthermore, slight increases in acinar cell adenomas of the pancreas and alveolar/bronchiolar adenomas in the lung were found in the males.

Chlorendic acid tested in $B6C3F_1$ mice fed diets containing 620 and 1250 mg/kg diet (equivalent to 62 and 125 mg/kg body weight) showed an increased incidence of necrosis and mitotic alteration in the liver. An increase in the incidence of hepato-cellular adenomas and carcinomas was found in males at both dose levels. An increased incidence of alveolar/bronchiolar adenomas or carcinomas was found in females.

Studies were carried out to investigate the mechanisms of carcinogenesis using an initiation/promotion assay, the partial hepatectomy model and the neonatal model. The tests showed that chlorendic acid has promoting activity.

### 1.1.7 Effects on humans

No data concerning effects on humans are available.

### 1.1.8 Effects on other organisms in the laboratory and field

Chlorendic acid has been reported to exert toxic effects on algae at 250 mg/litre. Effects reported include inhibition of microfaunal activity, decreased production of oxygen and decreased respiration. Toxic effects were not reported in algae exposed to 125 mg chlorendic acid/litre or in algae exposed to chlorendic anhydride. The toxic effects reported in algae have been attributed to pH change rather than direct toxicity of chlorendic acid. At low pH values, chlorendic acid exists in the non-ionized form, which may exert direct toxic effects.

Effects of chlorendic acid have been reported for a single species of terrestrial plant. Inhibition of both growth and seed germination was reported following exposure to 0.1 mg/litre or more, but no effects were reported following exposure to 0.01 mg/litre.

The 48-h $LC_{50}$ for *Daphnia magna* was 110.7 mg chlorendic anhydride/litre, and the 96-h $LC_{50}$ for both rainbow trout and bluegill sunfish was 422.7 mg/litre.

The potential effects of chlorendic anhydride on organisms in the environment cannot be evaluated without data on the concentrations and fate processes of this compound in environmental compartments.

## 1.2 Conclusions

The database on chlorendic acid and chlorendic anhydride is far from being complete. For several studies no full reports (only abstracts) were available to the Task Group. In particular, no data are available on the ultimate fate of the substances as such or in their reacted form, nor are data available on the bioaccumulation and biomagnification potential. Moreover, data on the exposure of humans and of organisms in the environment are lacking.

Both substances seem to have low acute and subacute oral toxicity, but they are dermal, eye and respiratory irritants. From the results of long-term toxicity/carcinogenicity studies with chlorendic acid on rats and mice, it is concluded that chlorendic acid induces tumours in rats and mice and is, therefore, considered to have a carcinogenic potential. However, a full hazard assessment for humans and the environment cannot be made in view of the lack of data.

The present database is inadequate to support the commercial use of chlorendic acid and anhydride.

## 1.3 Recommendations

### *1.3.1 Protection of human health and the environment*

a) Exposure of the general population to chlorendic acid, chlorendic anhydride and products derived from them should be minimized.

b) Chlorendic acid and anhydride must be used only in closed systems to prevent exposure to the vapour and dust. Workers producing and handling these substances should be properly trained in safety procedures. They should be protected from exposure with adequate engineering controls and appropriate industrial hygiene measures.

c) Disposal of chlorendic acid and anhydride and their waste products must be by methods which ensure that the general population cannot be exposed and that exposure of the environment is minimized.

### 1.3.2 Further research

a) The database should be completed with adequate data on:

   i) the ultimate fate of these substances as such or in their reacted form;

   ii) bioaccumulation and biomagnification potential;

   iii) human and environmental exposure data;

b) Studies should be carried out on the combustion products of materials prepared or treated with the acid or the anhydride, on their toxicity by inhalation, and on the potential of these combustion products for environmental contamination.

c) The concentrations of chlorendic acid and anhydride in environmental compartments should be measured, or at least predicted, using available models.

d) Data on the presence of chromosomal aberrations in a metaphase analysis study should be generated. Some *in vivo* mutagenicity data will be required before a full hazard assessment can be made.

e) A three-generation reproduction study should be conducted.

f) The carcinogenic mechanism for both substances should be clarified.

# 2. IDENTITY, PHYSICAL AND CHEMICAL PROPERTIES, AND ANALYTICAL METHODS

Chlorendic acid and chlorendic anhydride are closely related compounds. When chlorendic acid is heated in an open system it loses water and forms chlorendic anhydride. Chlorendic anhydride can be hydrolysed to chlorendic acid (Yu & Atallah, 1977a; Antonov, 1980; Larsen, 1980).

Exposure to chlorendic anhydride is likely to lead to exposure to chlorendic acid due to hydrolysis of the former (IARC, 1990).

## 2.1 Chlorendic acid

### 2.1.1 *Identity*

#### 2.1.1.1 *Primary constituent*

Chemical formula: $C_9 H_4 Cl_6 O_4$

Chemical structure:

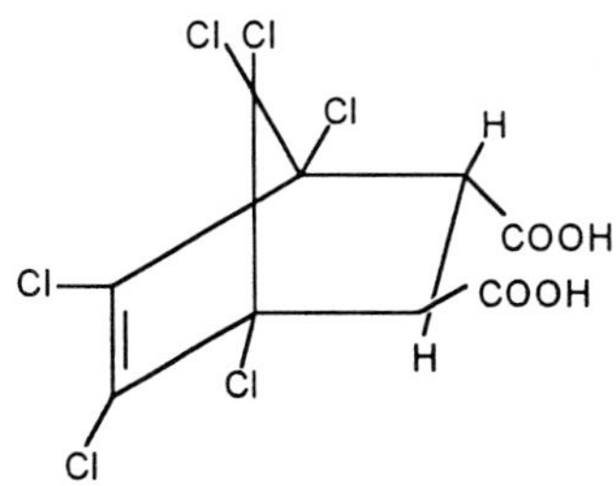


Chemical names:

- 1,4,5,6,7,7,-hexachloro-bicyclo[2,2,1]hept-5-ene-*endo*-*cis*-2,3-dicarboxylic acid (CAS);
- 1,4,5,6,7,7-hexachloro-5-norbornene-*endo*-*cis*-2,3-dicarboxylic acid (IUPAC);
- 1,4,5,6,7,7,-hexachloro-bicyclo [2,2,1]-5-heptene-2,3-dicarboxylic acid;
- hexachloro-*endo*-methylene-tetrahydrophthalic acid

| | |
|---|---|
| Common name: | chlorendic acid |
| Trade name: | HET acid |
| Relative molecular mass: | 388.87 |
| CAS number: | 115-28-6 |

#### 2.1.1.2 *Technical product*

Both chlorendic acid and chlorendic anhydride have similar chemical structure, using the same building blocks (hexachlorocyclopentadiene and maleic anhydride) in their production processes. Both products are manufactured in a closed system by a Diels-Alder reaction. The solvent used in the two processes differs. Chlorendic anhydride will form chlorendic acid in the presence of water (Yu & Atallah, 1977a; Occidental Chemical, 1988).

Chlorendic acid has only one commercial grade of high purity (99.5%). Common impurities are as follows:

| | |
|---|---|
| unreacted maleic anhydride | ≤ 0.25% |
| Fe | ≤ 1 ppm |
| hexachlorocyclopentadiene | ≤ 50 ppm |
| moisture | ≤ 1% |
| other volatiles | ≤ 0.25% |

(Occidental Chemical, 1988).

## 2.2 Chlorendic anhydride

### 2.2.1 *Identity*

#### 2.2.1.1 *Primary constituent*

Chemical formula: $C_9H_2Cl_6O_3$

Chemical structure:

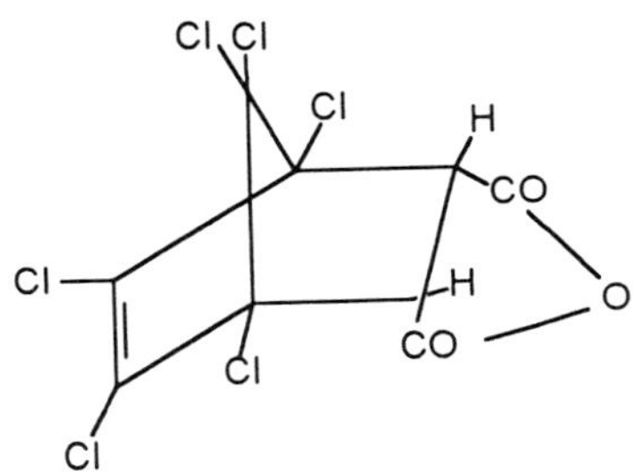

Chemical names:

- 4,5,6,7,8,8-hexachloro-3$^{a}$,4,7,7$^{a}$-tetrahydro-4,7-methano-isobenzofuran-1,3-dione (CAS);
- 1,4,5,6,7,7-hexachloro-*endo*-5-norbornene-2,3-dicarboxylic anhydride;
- hexachloro-*endo*-methylene tetrahydrophthalic anhydride;
- 1,4,5,6,7,7-hexachloro-*endo*-bicyclo[2,2,1]-hept-5-ene-2,3-dicarboxylic anhydride.

Common names: chlorendic anhydride

Trade name: HET anhydride (this name is no longer used)

Relative molecular mass: 370.86

CAS number: 115-27-5

RTECS number: RB 9080000

#### 2.2.1.2 *Technical product*

Chlorendic anhydride is available in two main grades "Refined" and "Technical", the latter being more commonly used.

Depending on the grade, the purity varies from 95% to 97%. Therefore, common impurities and volatile compounds are higher in chlorendic anhydride than in chlorendic acid (Velsicol Chemical Corp., 1982).

## 2.3 Physical and chemical properties

### 2.3.1 *Chlorendic acid*

Chlorendic acid is a hexachloro-norbornene compound, structurally related to the chlorinated cyclodiene insecticides such as aldrin, dieldrin, endrin, isodrin, endosulfan, chlordane and heptachlor. It is a fine white anhydrous non-dusting crystalline powder that is slightly soluble in water (0.35% by weight at 22.8 °C) and non-polar organic solvents (benzene (0.81%), carbon tetrachloride (0.21%), *n*-hexane (0.2%) and linseed oil (9.65%)) and is readily soluble in more polar organic solvents (e.g., methanol, ethanol and acetone). The chlorine content is 54.7% (US NTP, 1987; Occidental Chemical, 1988).

| | |
|---|---|
| Melting point: | 208-210 °C (sealed tube); 230-235 °C (open tube); conversion to the anhydride occurs prior to melting (Gupta et al., 1978; Occidental Chemical, 1987). |
| Stability: | The acid loses water in a heated open system, tends to discolour and forms the anhydride, which melts at 230-235 °C (Gupta et al., 1978). It emits chlorine when heated to decomposition at temperatures above 200 °C (Occidental Chemical, 1987). Chlorendic acid is very resistant to hydrolytic dechlorination, readily forms salts with a variety of metals, forms esters by heating with or without azeotropic solvent (e.g., chlorobenzene) and readily forms alkyl-type polyester resins by reaction with glycols and other polyols (Kirk-Othmer, 1981). |
| *n*-Octanol/water partition co-efficient (log $P_{oct/w}$): | 2.30 (Chemical Information Systems, 1988)<br>2.21 (Yu & Atallah, 1977b) |

### 2.3.2 Chlorendic anhydride

Chlorendic anhydride is a white crystalline substance with a melting point of 230-235 °C (US NTP, 1987). Commercial chlorendic anhydride contains 1-3% chlorendic acid (Velsicol Chemical Corp., 1982).

In aqueous solution, chlorendic anhydride is rapidly hydrolysed to chlorendic acid, with a half-life of approximately one hour (Yu & Atallah, 1977a).

Chlorendic anhydride will react almost quantitatively with excess methanol to form monomethyl chlorendate (Yu & Atallah, 1977a).

The partition coefficient for 1,2-dichlorobenzene[a]/water is 0.49 (Yu & Atallah, 1977b).

## 2.4 Analytical methods

Chlorendic acid may be determined by extraction with methanol and methylation with either 20% $BF_3$/methanol (Occidental Chemical, 1986) or diazomethane (Pilenkova & Fatyanova, 1980). The dimethyl ether is then determined by gas chromatography with electron capture detection after separation in packed columns (3% OV-1 on Chromosorb WHP, 80/100) or in capillary columns (Occidental Chemical, 1986).

The acid may also be analysed directly by $C_{18}$ reverse phase HPLC elution with water/acetonitrile/acetic acid and ultraviolet detection at 222 nm (Dietz et al., 1993). The detection limit for this method is 200 $\mu$g/litre with an average recovery of over 95%.

The earlier gas chromatography methods for direct analysis of the acid are not to be recommended.

The anhydride may be analysed by extraction and hydrolysis to the acid (Pilenkova & Fatyanova, 1980). The acid can then be analysed by any of the methods described above.

### *2.4.1 Air sampling*

Air concentrations of chlorendic acid and chlorendic anhydride can be measured by trapping the compounds on a glass-fibre filter. A portable pump is used to suck the air at a flow rate of 1 to 2 litre/min for 8 h. The samples are desorbed into methanol and can be analysed as described above (Occidental Chemical, 1986).

---

[a] 1,2-dichlorobenzene and *n*-octanol have similar dielectric constants; chlorendic anhydride would react with *n*-octanol to form a chlorendate.

# 3. SOURCES OF HUMAN AND ENVIRONMENTAL EXPOSURE

## 3.1 Natural occurrence

Chlorendic acid and chlorendic anhydride are not known to occur as natural products.

## 3.2 Anthropogenic sources

### *3.2.1 Production levels and processes*

Chlorendic acid and chlorendic anhydride production levels peaked in 1984 at an estimated annual volume of 6000 tonnes (US ITC, 1988). Since then, production levels have continuously declined, the 1993 estimated worldwide volume being 4000 tonnes.

Both products are prepared in a closed system, using the Diels-Alder reaction of hexachlorocyclopentadiene with maleic anhydride. The solvent used in the production process of the two products differs. The hydrolysis of the anhydride will lead to chlorendic acid having a higher purity than the anhydride form.

Chlorendic acid is currently manufactured in Belgium, and this is the sole production source in the world.

Chlorendic anhydride is produced only by one supplier located in the USA.

### *3.2.2 Uses*

Chlorendic acid and chlorendic anhydride are reactive chemical intermediates used primarily in the preparation of flame-retardant polyester resins and plasticizers. They are among the most reactive flame retardants in use. Both are used as chemical intermediates in the manufacture of corrosion-resistant (unsaturated) polyester resins with special applications in electrical systems, panelling, engineering, plastics and paints (Makhoulf, 1982). A major market is in fibreglass-reinforced resins for process equipment in the chemical industry. Chlorendic acid is also used to impart flame resistance to polyurethane foams when reacted with non-halogenated glycols to form halogenated polyols. In addition, it can be used in the manufacture of alkyl resins for special paints and inks. As an additive, it is used in acrylonitrile-butadiene-styrene copolymer (ABS) and polypropylene, but only in very limited amounts (Gupta et al., 1978; Larsen, 1980; Talbot, 1984; Occidental Chemical, 1988).

The double bond in chlorendic acid is not reactive as a cross-linkage site. Hence, reactive chemicals, such as maleic anhydride, glycols and styrene monomers, must be included in the polyester backbone to achieve cross-linkage (Gupta et al., 1978; Larsen, 1980; Talbot, 1984).

In Europe, 80% of chlorendic acid is used in composites for the building or transportation market where flame retardancy is required. The remainder is used in composites for the manufacture of anti-corrosion equipment such as tanks, piping and scrubbers. In the USA, Latin America and S.E. Asia, the usage pattern is reversed: 70-80% is for the anti-corrosion market and 20-30% is for flame retardant applications (Occidental Chemical, 1988).

### 3.2.3 Contamination of the environment

Chlorendic acid may be released, via hydrolytic degradation of polyesters, to the environment (soil and water) after their disposal (US NTP, 1987).

Chlorendic acid is also an oxidation product of endosulfan, chlordane, heptachlor, aldrin, dieldrin, isodrin and endrin, and their metabolites (Martens, 1972; Cochrane & Forbes, 1974; Menzie, 1978); it could, therefore, appear in the environment from sources other than direct fugitive emissions. It may also enter the environment after oxidation of other hexachlorocyclopentadiene-derived products.

Chlorendic acid may also enter the aquatic environment in wastewater from flame-proofing processes in the textile industry (Friedman et al., 1973, 1974).

# 4. ENVIRONMENTAL TRANSPORT, DISTRIBUTION AND TRANSFORMATION

## 4.1 Transport and distribution between media

No data on transport and distribution between media are available.

## 4.2 Transformation

### *4.2.1 Biodegradation*

Studies on *Clostridium butyricum* indicate that chlorendic acid is resistant to hydrolytic dechlorination and that it is not easily degradable (Schuphan & Ballschmiter, 1972).

### *4.2.2 Abiotic degradation*

#### *4.2.2.1 Chlorendic acid*

From a comparison of the characteristics of the cyclodiene derivatives, chlorendic acid would be expected, on theoretical grounds, to degrade by direct photolysis or by reactions with hydroxyl radicals and ozone (Parlar & Korte, 1977).

An experimental study was conducted to investigate the use of ozone to dechlorinate chlorendic acid. The dechlorination and subsequent degradation of chlorendic acid by ozonation was influenced by the pH, applied ozone dose and bicarbonate concentration. A change in the initial chlorendic acid concentration to 50, 100 and 200 mg/litre did not influence the rate of degradation of chlorendic acid. Ultraviolet (UV) radiation alone dechlorinates chlorendic acid. UV radiation was also shown to greatly enhance the oxidation of chlorendic acid in the presence of ozone. In a typical case, 80% dechlorination of chlorendic acid was obtained in 60 min when using an ozone dose of 125 mg/min ozone at pH 7.4. Conditions favouring radicals in solution, such as high pH and exposure to UV light, resulted in much faster dechlorination. The conditions which did not favour radicals, such as low pH and the addition of bicarbonate, resulted in slower dechlorination (Stowell & Jensen, 1991).

The photolysis of chlorendic acid by UV light and sunlight has been determined on solid surface and in aqueous solution. On a solid surface, chlorendic acid was degraded by UV light (half-life

= 16 days) to a number of unknown products. The irradiation of chlorendic acid in aqueous solution with UV light showed that the half-life in this system was 5 days (Yu & Atallah, 1978).

In soil, the half-life of $^{14}C$-chlorendic acid was found to be 140 ± 37 days at a soil concentration of 1 mg/kg and 280 ± 35 days at 10 mg/kg. However, it should be noted that the labelled carbon atoms were those in the chlorinated bicycloheptene ring and that these half-lives are more representative of that moiety than of chlorendic acid itself. Chlorendic acid can thus be considered fairly persistent in soil (Butz & Atallah, 1979a).

#### 4.2.2.2 Chlorendic anhydride

In aqueous solution, chlorendic anhydride is rapidly hydrolysed to chlorendic acid, with a half-life of approximately one hour (Yu & Atallah, 1977a).

The photolysis of chlorendic anhydride by UV light and sunlight has been investigated. On a solid surface, chlorendic anhydride rapidly absorbed water and was hydrolysed to chlorendic acid. Irradiation of chlorendic anhydride with sunlight on a thin-layer plate resulted in the formation of chlorendic acid and an unknown compound (Yu & Atallah, 1978).

### 4.2.3 Bioaccumulation and biomagnification

The octanol/water partition coefficient of chlorendic acid was found to be 2.21, and the 1,2-dichlorobenzene/water partition coefficient for chlorendic anhydride was found to be 0.49. Chlorendic anhydride rapidly converts to chlorendic acid and, as a result, will not bioconcentrate (Yu & Atallah, 1977a). The results indicate that neither chemical is likely to bioconcentrate in the environment (Yu & Atallah, 1977a,b).

No actual measured levels are available.

## 4.3 Ultimate fate following use

No data on ultimate fate following use are available.

# 5. ENVIRONMENTAL LEVELS AND HUMAN EXPOSURE

Chlorendic acid has been found in landfill leachate in the USA at concentrations up to 455 mg/litre. After treatment by powdered activated charcoal in sequencing batch reactors, the concentration of chlorendic acid was reduced to 51 mg/litre in the effluent (Ying et al., 1987).

# 6. KINETICS AND METABOLISM IN LABORATORY ANIMALS

## 6.1 Chlorendic acid

### 6.1.1 *Absorption, distribution and elimination*

#### 6.1.1.1 *Oral administration*

After oral administration of $^{14}C$-chlorendic acid (99%), in a solution of "Emulphor", ethanol and water, at 3 mg/kg body weight to male F-344/N rats, the compound was rapidly distributed, metabolized and eliminated. Approximately 78% of a single oral dose was excreted as conjugates in faeces within 24 h. The conjugates were resistant to beta-glucuronidase or arylsulfatase. Biliary excretion was the primary route of removal of the radioactivity from the liver. Only 3-6% was eliminated in the urine (Decad & Fields, 1982).

$^{14}C$-Chlorendic acid was administered orally (0.5-56 mg/kg body weight) to male rats to study the distribution and elimination. A mean total of 90% of the dose was recovered in excreta over 48 h and no major changes in the routes of elimination were observed over the dose range investigated. The major route of elimination was via faeces (87%), while 3% was excreted in urine. Expired air accounted for less than 0.5% of the radioactivity. After 48 h, the radioactivity in the liver increased in proportion to the dose, but there was no evidence of accumulation over the dose range studied. In general the concentrations in the other organs and tissues showed the same tendency, but the concentrations of radioactivity were about 4 to 5 times lower in the kidneys, whole blood and plasma, and lung than they were in the liver. The other organs contained at least 10 times lower concentrations.

Liver extracts of the animals exposed to 3 and 56 mg/kg showed unchanged chlorendic acid and at least one unidentified metabolite (Gurba et al., 1990).

#### 6.1.1.2 *Intravenous administration*

After intravenous administration of $^{14}C$-chlorendic acid (99%) in a solution of "Emulphor", ethanol and water, at 3 mg/kg body weight to male F-344 rats, the compound was rapidly distributed, metabolized and eliminated. The main route of elimination was via the faeces. The liver, blood, muscle, skin and kidneys were the most important depots for chlorendic acid, especially during the first hours. More than 50% of the total dose was found in the liver within 15 min. Radioactivity was rapidly removed from this organ and, by 7 h after the administration, less than 4% was present. Biliary excretion was the primary route of removal of the radioactivity from the liver. This decrease followed a single-component exponential curve. The half-life of chlorendic-acid-derived radioactivity from liver into bile was 1.19 h. The half-life for blood was 0.84 h, for muscle tissue 0.57 h and for skin 0.6 h. Other organs had only low levels of radioactivity, except for the adrenals. These organs had a greater specific activity than the liver for the first 3 h (Decad & Fields, 1982).

## 6.2 Chlorendic anhydride

### *6.2.1 Absorption, distribution and elimination*

The pharmacokinetics of chlorendic anhydride was evaluated in four male and eight female Holzman's albino rats that ingested $^{14}C$-chlorendic anhydride via gavage in a single dose of 3.65 (Group 1: 2 females), 4.00 (Group 2: 4 females), 5.55 (Group 3: 2 females) or 3.62 mg/kg (Group 4: 4 males). Blood was drawn from females treated with 4 mg/kg at 1, 4, 8, 17, 24, 48, 72 and 96 h after dosing. Urine and faeces were collected daily in the Group 1, 3 and 4 rats. Animals were sacrificed 17, 96, 192 and 192 h, respectively, after treatment, and selected tissues were excised. Regardless of sex, the primary route of excretion was the faeces, 70% of the administered dose being eliminated within the first 72 h. Only 10% of the administered dose was eliminated in the urine. After 192 h, the maximum residues present in any tissue were less than 0.1 mg/kg, with the exception of the fat (0.121 mg/kg) and liver (0.296 mg/kg). The blood concentration of chlorendic anhydride peaked one hour after dosing and was significantly decreased by 96 h. The absorption of chlorendic anhydride followed the two-compartment open model. The first compartment was suggested to consist of the blood and selected tissues which equilibrated rapidly, while the second compartment consisted of the fat which was slow in equilibrating and considered a deep reservoir. The half-life of the radiocarbon was less than 2 days except in fat (22.5 days) (Diaz & Atallah, 1978).

# 7. EFFECTS ON LABORATORY MAMMALS AND *IN VITRO* TEST SYSTEMS

## 7.1 Single exposure

### *7.1.1 Oral exposure*

#### *7.1.1.1 Chlorendic acid*

The acute oral $LD_{50}$ of chlorendic acid in rats (strain unspecified) was found to be 1770 mg/kg body weight (US NTP, 1987). The acute oral toxicity was evaluated in Charles River-CD male rats (number not reported) administered single doses of chlorendic acid, as a solution in acetone-peanut oil (1+9), by oral gavage at levels of 1.5, 2, 12, 130, 670, 1000, 1500, 2250, 3400 and 5000 mg/kg body weight. Mortality was observed at 2250 mg/kg or more. Clinical observations included discomfort, inactivity, irregular respiration, and weight loss. Gross necropsy was not reported (personal communication from E.I. du Pont de Nemours and Co. to the IPCS).

#### *7.1.1.2 Chlorendic anhydride*

The acute oral $LD_{50}$ of crude chlorendic anhydride in albino rats was found to be 2480 mg/kg body weight (Trzyna & Paa, 1975).

### *7.1.2 Dermal exposure*

#### *7.1.2.1 Chlorendic anhydride*

The acute dermal $LD_{50}$ of crude chlorendic anhydride in albino rabbits was found to be > 3000 mg/kg body weight (Trzyna & Paa, 1975).

### *7.1.3 Inhalation exposure*

#### *7.1.3.1 Chlorendic acid*

The acute inhalation toxicity of chlorendic acid was evaluated in groups of six male Charles River-CD rats exposed to chlorendic acid at chloride concentrations of 0.064, 0.066 or 0.102 mg/litre under nitrogen, and at chloride concentrations of 0.065 or 0.095 mg/litre under air for 4 h. The test atmosphere was generated by passing metered houseline air or nitrogen through a three-neck flask containing the test material heated to 100 °C. Mortality was observed in one animal in the 0.066-mg/litre group, all six in the 0.102-mg/litre group, three in the 0.065-mg/litre group, and all six in the 0.095-mg/litre group. The approximate lethal concentration was reported to be 0.066 mg/litre under nitrogen and 0.095 mg/litre under air. Clinical observations included lacrimation, irregular breathing, paleness, gasping, congestion, bloody nose, hyperaemia and weight loss. Histopathological examination revealed severe pulmonary irritation (personal communication from E.I. du Pont de Nemours and Co. to the IPCS).

An acute inhalation toxicity study was conducted with five female and five male Charles River-CD rats administered chlorendic acid dust at a maximum attainable concentration of 0.79 mg/litre of air in a 80-litre dynamic air flow chamber for 4 h. No deaths or toxic signs were observed during the exposure or during the 14-day observation period. The average 2-week body weight gains were within normal limits. Necropsy of animals did not reveal any gross pathological alterations (Kinert & Goode, 1975).

#### 7.1.3.2 *Chlorendic anhydride*

In a 4-h vapour inhalation study with crude chlorendic anhydride in albino rats, the acute $LC_{50}$ was found to be > 5.27 mg/litre. The observation period was 14 days (Myers & Goode, 1975).

An acute inhalation toxicity study was conducted with five male and five female Charles River CD rats given whole body exposure to the dust of chlorendic anhydride in a dynamic air flow chamber for 1 h. The chamber dust concentration was calculated to be 203 mg/litre. All animals survived the exposure and subsequent 14-day observation period. After 30 min of exposure, six rats exhibited salivation, and by the end of the exposure all animals exhibited salivation and one animal exhibited nasal discharge. After exposure, all animals appeared normal (Leong et al., 1978).

## 7.2 Short-term exposures

### 7.2.1 *Oral*

#### 7.2.1.1 *Mice*

Four or five male and five female B6C3F$_1$ mice (6-7 weeks old) were fed diets containing 0, 3100, 6200, 12 500, 25 000 or 50 000 mg/kg diet (equivalent to 0, 310, 620, 1250, 2500 or 5000 mg/kg body weight) chlorendic acid for 14 days. With the highest dose, the mice lost weight and four out of five males died. Mice receiving diets with 6200 mg/kg diet or more gained less weight than the controls. No treatment-related gross lesions were observed at necropsy (US NTP, 1987).

Four-week-old male and female B6C3F$_1$ mice were administered a diet containing 0, 1250, 2500, 5000, 10 000 or 20 000 mg/kg diet (equivalent to 0, 125, 250, 500, 1000 or 2000 mg/kg body weight) chlorendic acid for 13 weeks. Groups of 10 mice of each sex were used. Growth depression was seen in all groups, but with 5000 mg/kg diet or more this effect was significant. Feed consumption was not notably affected. No evidence of other compound-related effects was seen, except centrilobular hyperplasia, mitotic alterations and coagulative necrosis in the liver, especially in the animals given the two highest dose levels. The NOEL level was established at 250 mg/kg body weight (US NTP, 1987).

#### 7.2.1.2 *Rats*

*a) Chlorendic acid*

Five male and five female F-344/N rats (6-7 weeks old) were fed diets containing 0, 3100, 6200, 12 500, 25 000 or 50 000 mg/kg diet (equivalent to 0, 155, 310, 625, 1250 or 2500 mg/kg body weight) chlorendic acid for 14 days. With the highest dose, three out of five males and two out of five females died, and with 6200 mg/kg or more decreased growth was observed. No treatment-related gross observations were reported (US NTP, 1987).

Four-week-old male and female F-344/N rats (groups of 10 animals of each sex) were administered diets containing 0, 620, 1250, 2500, 5000 or 10 000 mg/kg diet (equivalent to 0, 31, 62.5,

125, 250 or 500 mg/kg body weight) chlorendic acid for 13 weeks. Growth depression was seen in both sexes with dose levels of 1250 mg/kg diet or more. In addition, the food intake was lower, especially in the groups fed with 2500 mg/kg or more. No abnormalities were seen except changes in the liver. The changes were hepatocytomegaly, mitosis alteration (increase in mitotic figures/field and abnormal mitotic figures) and bile duct hyperplasia in rats with 5000 and 10 000 mg/kg diet (equivalent to 250 and 500 mg/kg body weight, respectively) (US NTP, 1987).

*b) Chlorendic anhydride*

Four groups of Charles River CD rats, 15 rats of each sex per group, were fed diets containing 0, 100, 500 or 2500 mg/kg (equivalent to 0, 5, 25 or 125 mg/kg body weight per day) chlorendic anhydride for 90 days. They were observed twice daily. Haematological and biochemical tests and urinalyses were performed at 1, 2 and 3 months of the study for five rats of each sex per group. Three high-dosed females died between the fifth and thirteenth week of the study. Mid- and high-dose males and high-dose females had a decreased food consumption, and the mean body weight of mid- and high-dose males and all three treated female groups was decreased, compared to controls. Treated males and females had elevated serum alkaline phosphatase activities at 1, 2 and 3 months of the study. The mean absolute heart weight of male rats and the mean absolute and relative liver weight of male and female rats were statistically significantly decreased at all treatment levels. No compound-related gross lesions were seen in any of the treated rats, and no compound-related microscopic lesions were seen in 10 males and 10 females from the 2500-mg/kg group (Jefferson & Goldenthal, 1980).

### 7.2.2 Dermal

#### 7.2.2.1 Chlorendic anhydride

Four groups of New Zealand White rabbits (4 of each sex per group) were administered chlorendic anhydride at 0, 100, 500 or 2500 mg/kg body weight per day onto their clipped and/or abraded backs, 5 days/week for 3 weeks. The compound was wetted with physiological saline prior to dosing. All rabbits survived the treatment period. No compound-related changes were seen in urinalysis and haematological and biochemical investigations. All rabbits in the high-dose group had decreased body weight. Incidental findings, mainly in the high-dose group, were diarrhoea, nasal or ocular discharge, hypoactivity, anorexia and dehydration. One or more signs of dermal irritation, such as erythema, oedema, atonia, desquamation, coriaceousness and fissuring, were seen in all groups in a dose-related matter. Stomach mucosal erosions and ulcerations were found at necropsy in the two highest exposure groups. Apart from the skin irritation found at all dosages, the NOEL in this study was 100 mg/kg body weight per day (Goldenthal et al., 1979).

### 7.2.3 Inhalation

#### 7.2.3.1 Chlorendic anhydride

Four groups of Charles River CD rats (10 of each sex per group) were exposed to chlorendic anhydride dust for 6 h per day, 5 days/week during a 28-day experimental period. The average nominal exposure concentrations were 0, 0.11, 0.99 or 9.97 mg/litre. The median dust diameter was 6.0 (± 3.16) $\mu$m. All animals survived. All animals exhibited dose-related ocular and nasal irritation as well as salivation, following the daily exposures. They also exhibited a dose-related alopecia. Male rats at the highest dose level had decreased weight gains. Statistical differences were seen in haematocrit and erythrocyte values (males and females), haemoglobin (males), leukocytes (females), alkaline phosphatase levels (males and females) and glucose and serum glutamic pyruvic transaminase levels (males). Dark red foci and dark red discolouration in the lungs and dark red or brown foci in the glandular part of the stomach were seen at necropsy in the treated groups. Relative liver weights of males were decreased in all treated groups. The absolute and relative weight of the thyroid glands were decreased in females of the two highest dose groups. Compound-related microscopic changes of a haemorrhagic inflammatory nature in the lungs and of an inflammatory nature in the trachea, nasal turbinates and stomach mucosa occurred in rats from all treated groups. A NOEL was not established in this study (Ulrich, 1980).

## 7.3 Long-term exposure

Data from long-term exposure studies are given in section 7.7.

## 7.4 Skin and eye irritation; sensitization

### 7.4.1 Chlorendic acid

Repeated application of chlorendic acid (as a powder or as a solution in dimethyl phthalate) to the skin of New Zealand rabbits caused local skin irritation (Witherup et al., 1965).

Repeated application of chlorendic acid (as a powder or as a 5, 10 or 20% solution in dimethyl phthalate) beneath the eyelids of New Zealand rabbits caused severe eye irritation (Witherup et al., 1965).

In a skin sensitization test in albino guinea-pigs, chlorendic acid gave a negative response (Brett, 1975).

### 7.4.2 Chlorendic anhydride

In a 3-week dermal toxicity study with chlorendic anhydride in New Zealand White rabbits, it was considered to be a mild skin irritant (see section 7.2.2) (Goldenthal et al., 1979).

Chlorendic anhydride was severely irritating to the eyes of albino rabbits (Trzyna & Paa, 1975).

In a dermal sensitization study in albino guinea-pigs, chlorendic anhydride produced a positive response, and it has to be considered as a possible dermal sensitizing agent in humans (Dean & Jessup, 1978).

## 7.5 Reproductive toxicity, embryotoxicity and teratogenicity

### 7.5.1 Chlorendic anhydride

The teratogenicity of chlorendic anhydride was evaluated in pregnant Charles River CD rats (25 per group) exposed orally by gavage at dose levels of 0, 25, 100 or 400 mg/kg chlorendic anhydride per day on gestational days 6-15. Cesarean sections were performed on day 20. Significant differences were observed between treated and control animals in the following: decreased maternal body weight and weight gain (high-dose group), fetal sex ratio (low-dose level), and increased mean number of post-implantation losses (mid- and high-dose groups). No significant differences were observed between treated and control animals in the following: mean number of *corpora lutea*, viable or non-viable fetuses, mean fetal body weights and fetal malformations (Goldenthal et al., 1978).

## 7.6 Mutagenicity and related end-points

### *7.6.1 Chlorendic acid*

Chlorendic acid was not mutagenic in *Salmonella typhimurium* strains TA100, TA98, TA1535 and TA1537, in the presence or absence of an exogenous metabolism system (liver S9 mix induced by Aroclor 1254 in male Sprague-Dawley rats or male Syrian hamsters). Dose levels tested were 100 to 7690 μg/plate (Haworth et al., 1983; Zeiger, 1990).

The mutagenicity of chlorendic acid was evaluated in *S. typhimurium* strains TA1537, TA1538, TA98, TA1535 and TA100, both in the presence and absence of metabolic activation by rat liver S9 fraction (inducer not reported). Following preliminary toxicity tests, assays using activation were performed at concentrations of 500, 1000, 1500, 2500, 5000 and 7500 μg/plate and with no activation at 50, 100, 250, 500 and 750 μg/plate. The test material did not produce a positive mutagenic response in any bacterial strain, either with or without activation (Dupont-deNemours, proprietary information (US EPA, 1982).

Chlorendic acid was tested in the Microscreen prophage-induction assay in *Escherichia coli* at seven dose levels ranging from 0.4 to 25.7 mM, with and without metabolic activation. In the repeat study only, a dubious positive reaction was found at the highest dose level with activation. In the first test this dose level was toxic (DeMarini & Brooks, 1992).

Chlorendic acid was mutagenic in the L5178Y/TK+/- mouse lymphoma assay in the absence of S9 activation. The assay was performed twice with the same results. Dose levels were 1300 to 1700 μg/ml. The highest (cytotoxic) dose showed a positive effect; an increase of total mutant clones, relative total growth depression and increase in mutation frequency (US NTP, 1987). Similar finding were reported by McGregor et al., (1988).

Chlorendic acid was positive in a standard transformation assay using BALB/c-3T3 cells without exogenous activation at concentrations of 2-4 mM (Matthews et al., 1993).

Chlorendic acid was tested for its ability to induce sex-linked recessive lethal mutations in post-meiotic and meiotic germ cells of male *Drosophila melanogaster*. Chlorendic acid was negative at 2000 and 15 000 mg/kg administered by injection (Foureman et al., 1994).

### 7.6.2 Chlorendic anhydride

The mutagenicity of chlorendic anhydride was evaluated in *Salmonella* tester strains TA98, TA100, TA1535, TA1537 and TA1538, and the yeast *Saccharomyces cerevisiae* tester strain D4, both in the presence and absence of metabolic activation by Aroclor-induced rat liver S9 fraction. Chlorendic anhydride dimethylsulfoxide (DMSO), at doses up to 500 μg/plate, did not cause a positive response in any of the bacterial strains or the yeast strain, either with or without metabolic activation (Jagannath & Brusick, 1977).

Chlorendic anhydride did not induce forward mutations at the TK locus in L5178Y mouse lymphoma cells at concentrations up to 0.24 mg/ml without, and up to 0.32 mg/ml with, an S9 activation system. Toxicity was observed with higher concentrations (Matheson & Brusick, 1978a).

Chlorendic anhydride was negative in an *in vitro* malignant transformation assay in BALB/3T3 cells without exogenous activation at concentrations of 0.005-0.078 mg/ml. Only at 0.010 mg/ml was a significant, but non-dose-related, increase in transformation frequency found (Matheson & Brusick, 1978b).

Chlorendic anhydride produced significant increases in unscheduled DNA synthesis assay in human WI-38 cells at dose levels up to 0.5 mg/ml (Matheson & Brusick, 1978c).

The mutagenicity of chlorendic anhydride was evaluated in a dominant lethal assay using four groups of 20 male CD-1 mice exposed orally by gavage at dose levels of 0, 22, 74 or 223 mg/kg in a single exposure (in DMSO vehicle). Following exposure, each male was rested for 2 days and then mated for 5 days/week with two untreated females each week for 7 consecutive weeks. Mated females were sacrificed 14 days after the midweek of the mating period. Pregnant females were scored for dominant lethal indices at mid-pregnancy. A statistically significant decrease was observed in the fertility index, relative to controls, for all females mated to treated males during week 5 and females mated to mid-dose level males during week 4. There were no differences between females mated to treated and control males with respect to average number of implantations per pregnant female, average resorptions per pregnant female, proportions of dead implants/implants, proportions of females with one or more resorptions, or proportions of females with two or more resorptions (Matheson & Brusick, 1978d). Although the study was reported as giving negative results, there were some statistically non-significant increases in dead implants per pregnant mouse in females mated at weeks 2 and 8 at the high dose and at week 5 in the low- and mid-dose treated mice. However, the study was found to be inadequately designed.

## 7.7 Carcinogenicity

### *7.7.1 Chlorendic acid*

#### *7.7.1.1 Mice*

Diets containing 0, 620 or 1250 mg chlorendic acid (> 98%)/kg diet (equivalent to 0, 62 and 125 mg/kg body weight per day) were fed to groups of 50 male and 50 female $B6C3F_1$ mice for 103 weeks. The estimated daily intake of chlorendic acid was 89 and 185 mg/kg body weight for low- and high-dose males and 100 and 207 mg/kg body weight for low- and high-dose female mice. All survivors were killed at week 112. The mice were 8 weeks old when placed on the diets. Survival and feed consumption of treated mice of both sexes were similar to those of controls, although mean body weights of high-dose males and females were lower than those of the controls. In the liver an increased incidence of necrosis was observed in dosed male mice (coagulative necrosis), and mitotic alterations were found in high-dose female mice. The incidence of hepatocellular adenomas was increased in males (see Table 1; controls 5/50 (10%), low-dose 9/49 (18%) and high-dose 10/50 (20%) (statistically significant)) and so was the incidence of hepatocellular carcinomas (male controls 9/50 (18%), low-dose 17/49 (35%) and high-dose 20/50 (40%) (statistically significant)). Hepatocellular carcinomas metastasized to the lung in males (controls 2/50 (4%), low-dose 4/49 (8%) and high-dose 7/50 (14%)). The incidence of alveolar/bronchiolar adenomas and carcinomas (combined) in females was controls 1/50 (2%), low-dose 5/50 (10%) and high-dose 6/50 (12%). However, the incidence in the concurrent

Table 1. Incidence of tumours in B6C3F$_1$ mice (from: US NTP, 1987)

| Type of tumour | Males | | | Females | | |
|---|---|---|---|---|---|---|
| | Control | 620 mg/kg diet | 1250 mg/kg diet | Control | 620 mg/kg diet | 1250 mg/kg diet |
| **Liver** | | | | | | |
| Hepatocellular adenoma | 5/50 (10%) | 9/49 (18%) | 10/50 (20%) | | | |
| Hepatocellular carcinoma | 9/50 (18%) | 17/49 (35%) | 20/50 (40%) | | | |
| Hepatocellular adenoma or carcinoma | 13/50 (26%) | 23/49 (47%) | 27/50 (54%) | 3/50 (6%) | 7/49 (14%) | 7/50 (14%) |
| **Lung** | | | | | | |
| Alveolar/bronchiolar adenoma or carcinoma | | | | 1/50 (2%) | 5/50 (10%) | 6/50 (12%) |
| **Thyroid** | | | | | | |
| Follicular cell adenoma | 0/50 (0%) | 0/47 (0%) | 3/50 (6%) | | | |

controls compared with the historical control average was low and the biological significance of these results is unclear. On the basis of the results from this study, it was concluded that there was clear evidence of carcinogenic potential of chlorendic acid in male $B6C3F_1$ mice, as demonstrated by an increase in hepatocellular carcinomas (US NTP, 1987; Huff et al., 1989; IARC, 1990).

#### *7.7.1.2 Rats*

Diets containing 0, 620 or 1250 mg chlorendic acid (> 98%)/kg diet (equivalent to 0, 31 and 62.5 mg/kg body weight per day) were fed to groups of 50 male and 50 female F-344/N rats for 103 weeks. The estimated daily intake of chlorendic acid was 27 and 56 mg/kg body weight for low- and high-dose males, and 39 and 66 mg/kg body weight for low- and high-dose female rats. All survivors were killed in week 112. The rats were 8 weeks old when placed on the diets. Survival and feed consumption of treated rats were similar to those of the controls, although mean body weight of high-dose males and females were lower. Significant changes were observed in the incidence of rats with neoplastic or non-neoplastic lesions of the liver, pancreas, lung, preputial gland, uterus, salivary gland, urinary system, mammary gland, adrenals, testes and pituitary gland. In the liver of male rats, increases in the incidence of cystic degeneration and focal cellular changes were found. An increase in the incidence of granulomatous inflammation was observed in females. Bile duct hyperplasia incidence was increased in both sexes.

From Table 2, it is clear that an increase in tumour incidence was found in the liver, pancreas, preputial gland and lung. The incidence of neoplastic nodules of the liver (adenomas) was significantly increased in treated males (controls 2/50 (4%), low-dose 21/50 (42%) and high-dose 23/50 (46%) (statistically significant)) and in females (controls 1/50 (2%), low-dose 3/49 (6%) and high-dose 11/50 (22%) (statistically significant)). The values for hepatocellular carcinomas in females were controls 0/50 (0%), low-dose 3/49 (6%) and high-dose 5/50 (10%) (statistically significant). Those for the acinar cell adenomas of the pancreas in males were controls 0/49 (0%), low-dose 4/50 (8%) and high-dose 6/50 (12%). The incidence of alveolar/bronchiolar adenomas was significantly increased in high-dose males (controls 0/50 (0%), low-dose 3/50 (6%) and high-dose 5/50 (10%)) and that of carcinomas of the preputial gland in males (controls 1/50 (2%), low-dose 8/50 (16%) and high-dose 4/50 (8%)). Slight increases

Table 2. Incidence of tumours in F-344/N rats (from: US NTP, 1987)

| Type of tumour | Males | | | Females | | |
|---|---|---|---|---|---|---|
| | Control | 620 mg/kg diet | 1250 mg/kg diet | Control | 620 mg/kg diet | 1250 mg/kg diet |
| **Liver** | | | | | | |
| Hepatocellular adenoma | 2/50 (4%) | 21/50 (42%) | 23/50 (46%) | 1/50 (2%) | 3/49 (6%) | 11/50 (22%) |
| Hepatocellular carcinoma | | | | 0/50 (0%) | 3/49 (6%) | 5/50 (10%) |
| **Pancreas** | | | | | | |
| Acinar cell adenoma | 0/49 (0%) | 4/50 (8%) | 6/50 (12%) | | | |
| **Lungs** | | | | | | |
| Alveolar/bronchiolar adenoma | 0/50 (0%) | 3/50 (6%) | 5/50 (10%) | | | |
| **Preputial gland** | | | | | | |
| Squamous cell papilloma, adenoma or carcinoma | 1/50 (2%) | 8/50 (16%) | 4/50 (8%) | | | |
| **Uterus/endometrium** | | | | | | |
| Endometrial stromal polyp | | | | 5/50 (10%) | 15/49 (31%) | 10/50 (20%) |
| **Salivery gland** | | | | | | |
| Fibrosarcoma/ neurofibrosarcoma | 1/50 (2%) | 2/49 (4%) | 4/50 (8%) | | | |

in the incidences of salivary gland tumours in males and endometrial stromal polyps in females were noted. Under the conditions of the experiment, it was concluded that chlorendic acid showed carcinogenic properties in F-344/N rats, demonstrated by increased incidence of adenomas of the liver in both sexes and of hepatocellular carcinomas in females. An increase in the incidence of acinar cell adenomas of the pancreas was found in males (US NTP, 1987; Huff et al., 1989; IARC, 1990).

In an initiation-promotion assay in rat liver designed as a complement to the 2-year bioassay, groups of male and female Fischer-344 rats were given a single oral dose of 10 mg/kg body weight diethylnitrosamine followed 24 h later by 70% partial hepatectomy. After a 2-week recovery period, animals were fed a basal diet containing 0, 620 or 1250 mg/kg diet (equivalent to 0, 31 and 62.5 mg/kg body weight per day) chlorendic acid for 6 months, at which time all surviving animals were killed. Altered hepatic foci (AHF) were measured using four markers: placental glutathione-*S*-transferase (PGST), $\gamma$-glutamyl transpeptidase, adenosine triphosphate and glucose-6-phosphatase. The number of AHF/liver and the volume percentage of liver as AHF were significantly increased by chlorendic acid at both dose levels. Placental GST was the best single marker for hepatic promoting activity in both sexes (Dragan et al., 1991).

Partial hepatectomy and neonatal rat short-term liver focus models were used to examine the potential of chlorendic acid for initiating and promoting activity. Chlorendic acid showed clear evidence of hepatocarcinogenicity. While it showed no initiating activity, promoting activity, as indicated by increased number, size or volume fraction of histochemically detected hepatic foci of cellular alteration, was evident (Maronpot et al., 1989).

#### 7.7.1.3 *Special studies*

Chlorendic acid was administered in single doses of 0, 450 or 900 mg/kg body weight to 9-week-old male F-344 rats by oral application or subcutaneous injection. Replicative DNA synthesis was measured in primary hepatocyte *in vitro* cultures prepared from the livers of the treated animals at 24, 39 or 48 h after treatment. Chlorendic acid did not increase replicative DNA synthesis in this test (Uno et al., 1994).

Nine female Sprague-Dawley (CD) rats (90-days old; 200-230 g) were administered two doses of chlorendic acid (159 mg/kg body weight) in corn oil by gavage, 21 and 4 h before sacrifice, and compared with corn oil controls. Hepatic DNA damage was indicated as an increased hepatic ornithine decarboxylase activity. However, there was no DNA damage (Kitchin et al., 1993).

### 7.7.2 Chlorendic anhydride

Chlorendic anhydride rapidly converts to chlorendic acid on contact with water.

A separate carcinogenicity study with chlorendic anhydride has not been conducted and seems not to be necessary, since chlorendic anhydride is metabolically transformed into chlorendic acid.

# 8. EFFECTS ON HUMANS

No data concerning the effects of chlorendic acid or anhydride on humans are available.

# 9. EFFECTS ON OTHER ORGANISMS IN THE LABORATORY AND FIELD

## 9.1 Chlorendic acid

Studies were conducted to estimate the effects of 125, 250 and 500 mg/litre chlorendic acid (99.5%) on an aquatic microcosm. To account for the pH effect of the acid, two additional flask experiments were conducted in which the solutions of chlorendic acid were titrated with 50% (w/w) NaOH to pH 7.4 or with HCl to create a working solution of pH 6.2, 4.6 and 3.5 (equivalent to the initial pH of the 125, 250 and 500 mg/litre chlorendic acid treatments, respectively). The experiments were carried out with naturally derived laboratory microcosms containing bacteria, fungi, unicellular and colonial green algae, filamentous blue-green algae, diatoms, protozoans, rotifers, copepods and ostracods. In 6-day flask studies, chlorendic acid concentrations of 500 mg/litre (pH 3.5) completely inhibited algal growth and microfaunal activity, 250 mg/litre (pH 4.1) inhibited microfaunal activity and reduced the abundance of all but one algal species, and 125 mg/litre (pH 6.2) had no observable effects. Similar results occurred in longer term microcosm studies where, in addition, 500 and 250 mg/litre chlorendic acid (no pH specified) resulted in decreased oxygen production and respiration, altered chlorophyll *a* concentrations and bacterial populations, and increased concentrations of dissolved $NO_3$-N, $NH_3$-N and $PO_4$-P. In contrast, few distinct effects were observed in flasks or microcosms treated with the non-ionized form of chlorendic acid. Results indicate that the observed effects at lower pH values were due primarily to increases in hydrogen ion concentration; direct toxicity also may have occurred at low pH, where chlorendic acid existed as the non-ionized species (Hendrix et al., 1983).

Chlorendic acid caused a dose-related inhibition of seed germination and early growth of garden cress (*Lepidium sativum* L). It inhibited growth by 0, 20, 60 and approximately 90% during exposure for 7 days to 0, 0.1, 1.0 and 10.0 mg/litre (concentrations of chlorendic acid in aquatic test solutions), respectively. At 0.001 and 0.01 mg/litre, no influence was found (Koch, 1970).

Butz & Atallah (1979b) studied the effect of chlorendic acid on soil microorganisms. A silty clay soil was treated with 0, 1 or 10 mg/kg chlorendic acid in Erlenmeyer flasks and incubated at 30 °C for 28 days. A significant increase in the number of soil fungi was reported for the 10-mg/kg group, but no effects on soil fungi were reported at 1 mg/kg. A non-significant increase in the number of bacteria was reported for both treated soils.

## 9.2 Chlorendic anhydride

The 48-h $LC_{50}$ of chlorendic anhydride for *Daphnia magna* is 110.7 mg/litre; the 48-h NOEL is 56 mg/litre (Vilkas & Hutchinson, 1977).

The 96-h $LC_{50}$ of chlorendic anhydride for rainbow trout (*Salmo gairdneri*) is 422.7 mg/litre nominal concentration in a static system. Above 100 mg/litre the trout become excitable (Calmbacher et al., 1977a).

The 96-h $LC_{50}$ of chlorendic acid for bluegill sunfish (*Lepomis macrochirus*) is 422.7 mg/litre (nominal concentration) in a static system. Below 320 mg/litre no abnormal behaviour is observed (Calmbacher et al., 1977b).

# 10. PREVIOUS EVALUATIONS BY INTERNATIONAL BODIES

Chlorendic acid has been evaluated by the International Agency for Research on Cancer in its series of IARC Monographs on the Evaluation of Carcinogenic Risks to Humans. The evaluation was that there is sufficient evidence for the carcinogenicity of chlorendic acid in experimental animals. No data were available from human studies on the carcinogenicity of chlorendic acid. The overall evaluation was that chlorendic acid is possibly carcinogenic to humans (Group 2B) (IARC, 1989, 1990).

## REFERENCES


Antonov LT (1980) [Selection of structural materials in chemical engineering.] Khim Prom-st (Moscow), **3**: 167-168 (in Russian).

Brett B (1975) Skin sensitization test with five samples in albino guinea-pigs. Northbrook, Illinois, Industrial Biotest Laboratories Inc. (Proprietary report submitted to WHO by Velsicol Chemical Corporation, Chicago, USA).

Butz RG & Atallah YH (1979a) Persistence of chlorendic acid in soil at 1 and 10 μg/g. Chicago, Illinois, Velsicol Chemical Corporation (Proprietary report).

Butz RG & Atallah YH (1979b) Effects of chlorendic acid at 1 and 10 μg/g on soil microflora. Chicago, Illinois, Velsicol Chemical Corporation (Proprietary report).

Calmbacher CW, Vilkas AG, & Hutchinson CH (1977a) The acute toxicity of chlorendic anhydride to the rainbow trout, *salmo gairdneri* Richardson. Tarrytown, New York, Union Carbide Environmental Services (Proprietary report No. 11506-03-07a, submitted to WHO by Velsicol Chemical Corporation, Chicago, USA).

Calmbacher CW, Vilkas AG, & Hutchinson CH (1977b) The acute toxicity of chlorendic anhydride to the Bluegill sunfish, *Lepomis macrochirus* Rafinesque. Tarrytown, New York, Union Carbide Environmental Services (Proprietary report No. 11506-03-07b, submitted to WHO by Velsicol Chemical Corporation, Chicago, USA).

Chemical Information Systems (1988) Information system for hazardous organics in water (ISHOW). Baltimore, Maryland, Infrared Search System (IRSS), Mass Spectral Search System (MSSS).

Cochrane WP & Forbes MA (1974) Oxidation products of heptachlor and its metabolites - a chemical study. Chemosphere, **3**: 41-46.

Dean WP & Jessup DC (1978) Dermal sensitization study in the albino guinea-pig with chlorendic anhydride. Mattawan, Michigan, International Research and Development Corporation (Proprietary report No. 163-529 to Velsicol Chemical Corporation, Chicago, USA).

Decad GM & Fields MT (1982) Disposition and excretion of chlorendic acid in Fischer 344 rats. J Toxicol Environ Health, **9**(5-6): 911-920.

Demarini DM & Brooks HG (1992) Induction of prophage lambda by chlorinated organics: detection of some single-species/single site carcinogens. Environ Mol Mutagen, **19**: 98-111.

Diaz L & Atallah YH (1978) Pharmacokinetics of chlorendic anhydride in rats. Chicago, Illinois, Velsicol Chemical Corporation (Proprietary report No. 482348).

Dietz EA, Cortellucci NJ, & Singley KL (1993) Determination of benzoic acid, chlorobenzoic acids and chlorendic acid in water. J Liq Chromatogr, **16**(15): 3331-3347.

Dragan YP, Rizvi T, Xu Y-H, Hully JR, Bawa N, Campbell HA, Maronpot RR, & Pitot HC (1991) An initiation-promotion assay in rat liver as a potential complement to the 2-year carcinogenesis bioassay. Fundam Appl Toxicol, **16**(3): 525/547.

Foureman P, Mason JM, Valencia R, & Zimmering S (1994) Chemical mutagenesis testing in Drosophila: X. Results of 70 coded chemicals tested for the National Toxicology Program. Environ Mol Mutagen, **23**(3): 208-227.

Friedman M, Whitefield RE, & Tillin S (1973) Enhancement of the natural flame-resistance of wool. Text Res J, **43**: 212-217.

Friedman M, Ash JF, & Fox W (1974) Dye bath application of chlorendic acid for flame-resistant wool. Text Res J, **44**: 555-556.

Goldenthal EI, Jessup DC, & Rodwell DE (1978) Chlorendic anhydride, teratology study in rats. Mattawan, Michigan, International Research and Development Corporation (Proprietary report No. 163.535, submitted to WHO by Velsicol Chemical Corporation, Chicago, USA).

Goldenthal EI, Jessup DC, Geil RG, & Dean WP (1979) Chlorendic anhydride. Three-week dermal toxicity study in rabbits. Mattawan, Michigan, International Research and Development Corporation (Proprietary report No. 163.530, submitted to WHO by Velsicol Chemical Corporation, Chicago, USA).

Gupta SK, Krishman M, & Thampy RT (1978) Preparation of chlorendic acid based polyester resins. Ind J Text Res, **3**: 124-128.

Gurba PE, Smith LW, & Cameron BD (1990) Disposition of chlorendic acid in the rat. Toxicologist, **10**(1): 237 (Abstract No. 948).

Haworth S, Lawlor T, Mortelmans K, Speck W, & Zeiger E (1983) *Salmonella* mutagenicity test results for 250 chemicals. Environ Mutagen, **1**(suppl): 3-142.

Hendrix PF, Hamala JA, Langner CL, & Kollig HP (1983) Effects of chlorendic acid, a priority substance, on laboratory aquatic ecosystems. Chemosphere, **12**(7/8): 1083-1099.

Huff JA, Eustis SL, & Haseman JK (1989) Occurrence and relevance of chemically induced benign neoplasms in long-term carcinogenicity studies. Cancer Metastasis, **8**: 1-21.

IARC (1989) Some organic solvents, resin monomers and related compounds, pigments and occupational exposures in paint manufacture and painting. Lyon, International Agency for Research on Cancer, 535 pp (IARC Monographs on the Evaluation of Carcinogenic Risks to Humans, Volume 47).

IARC (1990) Chlorendic acid. In: Some flame retardants and textile chemicals, and exposures in the textile manufacturing industry. Lyon, International Agency for Research on Cancer, pp 45-53 (IARC Monographs on the Evaluation of Carcinogenic Risks to Humans, Volume 48).

IPCS (1989) Environmental Health Criteria 88: Polychlorinated dibenzo-*para*-dioxins and dibenzofurans. Geneva, World Health Organization, 409 pp.

IPCS (1994a) Environmental Health Criteria 162: Brominated diphenyl ethers. Geneva, World Health Organization, 347 pp.

IPCS (1994b) Environmental Health Criteria 152: Polybrominated biphenyls. Geneva, World Health Organization, 577 pp.

IPCS (1995a) Environmental Health Criteria 172: Tetrabromobisphenol-A and derivatives. Geneva, World Health Organization, 139 pp.

IPCS (1995b) Environmental Health Criteria 173: Tris- and bis(2,3-dibromopropyl) phosphate. Geneva, World Health Organization, 129 pp.

Jagannath DR & Brusick DJ (1977) Mutagenicity evaluation of chlorendic anhydride: Final report. Kensington, Maryland, Litton Bionetics Inc. (Proprietary report No. 20838, submitted to WHO by Velsicol Chemical Corporation, Chicago, USA).

Jefferson ND & Goldenthal EI (1980) Chlorendic anhydride: 90-day subacute renal toxicity study in rats. Mattawan, Michigan, International Research and Development Corporation (Proprietary report No. 163.533, submitted to WHO by Velsicol Chemical Corporation, Chicago, USA).

Kinert JC & Goode JW (1975) Acute dust inhalation study in rats with Het acid. Northbrook, Illinois, Industrial Biotest Laboratories Inc. (Proprietary report No. 663-06279 to Hooker Chemical Corporation, Niagara Falls, USA).

Kirk-Othmer (1981) Encyclopedia of chemical technology, 3rd ed. New York, John Wiley and Sons, vol 10, pp 387-389.

Kitchin KT, Brown JL, & Kulkarni AP (1993) Predicting rodent carcinogenicity of halogenated hydrocarbons by *in vivo* biochemical parameters. Teratogen Carcinogen Mutagen, **13**(4): 167-184.

Koch H (1970) [Phytotoxic substances and their possible mechanism of action.] Sci Pharm, **38**(2): 79-98 (in German).

Larsen ER (1980) Flame retardants: Phosphorous compounds. In: Mark HF, Othmer DF, Overberger CG, Seaborg GT, & Grayson M ed. Encyclopedia of chemical technology, 3rd ed. New York, John Wiley & Sons, vol 10, pp 388-389.

Leong KJ, Cavender FL, Branch JW, & Sabaitis CP (1978) Chlorendic anhydride: Acute inhalation toxicity study in rats. Mattawan, Michigan, International Research and Development Corporation (Proprietary report No. 163.512, submitted to WHO by Velsicol Chemical Corporation, Chicago, USA).

McGregor DB, Brown A, Cattanach P, Edwards I, McBride D, & Caspary WJ (1988) Response of the L5178Y tk+/tk- mouse lymphoma cell forward mutation assay.II: 18 Coded chemicals. Environ Mol Mutagen, **11**(1): 91-118.

Makhoulf J (1982) Polyesters unsaturated. In: Mark HF, Othmer DF, Overberger CG, Seaborg GT, & Grayson M ed. Encyclopedia of chemical technology, 3rd ed. New York, John Wiley & Sons, vol 18, pp 579-594.

Maronpot RR, Pitot HC, & Peraino C (1989) Use of rat liver altered focus models for testing chemicals that have completed two-year carcinogenicity studies. Toxicol Pathol, **17**(4): 651-662.

Martens R (1972) [Degradation of endosulfan by soil microorganisms.] Schr.reihe Ver Wasser Boden Lufthyg (Berlin-Dahlem), **37**: 167-173 (in German).

Matheson DW & Brusick DJ (1978a) Mutagenicity evaluation of chlorendic anhydride in the mouse lymphoma forward mutation assay. Kensington, Maryland, Litton Bionetics Inc. (Proprietary report No. 20839, submitted to WHO by Velsicol Chemical Corporation, Chicago, USA).

Matheson DW & Brusick DJ (1978b) Evaluation of chlorendic anhydride: *in vitro* malignant transformation in BALB/c3T3 cells. Kensington, Maryland, Litton Bionetics Inc. (Proprietary report, project no. 20840, submitted to WHO by Velsicol Chemical Corporation, Chicago, USA).

Matheson DW & Brusick DJ (1978c) Mutagenicity evaluation of chlorendic anhydride in the unscheduled DNA synthesis in human WI-38 cells assay. Kensington, Maryland, Litton Bionetics Inc. (Proprietary report, project no. 20840, submitted to WHO by Velsicol Chemical Corporation, Chicago, USA).

Matheson DW & Brusick DJ (1978d) Mutagenicity evaluation of chlorendic anhydride in the mouse dominant lethal assay. Kensington, Maryland, Litton Bionetics Inc. (Proprietary report No. 20862, submitted to WHO by Velsicol Chemical Corporation, Chicago, USA).

Mathews EJ, Spalding JW, & Tennant RW (1993) Transformation of BALB/c3T3 cells: V. Transformation responses of 168 chemicals compared with mutagenicity in *Salmonella* and carcinogenicity in rodent bioassays. Environ Health Perspect, **101**(suppl 2): 347-482.

Menzie CM (1978) Metabolism of pesticides: update II to special scientific report No. 212. Washington, DC, US Department of Interior, Fish and Wildlife Service, pp 131-133.

Myers TW & Googe JW (1975) Acute vapor inhalation toxicity study with crude Het anhydride. Northbrook, Illinois, Industrial Biotest Laboratories Inc. (Proprietary report No. 663-06667).

Occidental Chemical (1986) Analytical procedure for the determination of HET acid trapped on glass fibre filters. Brussels, Occidental Chemical.

Occidental Chemical (1987) Material safety data sheet M8584: HET acid. Niagara Falls, New York, Occidental Chemical.

Occidental Chemical (1988) HET acid. Brussels, Occidental Chemical.

Parlar H & Korte F (1977) Photoreactions of cyclodiene insecticides under simulated environmental conditions - A review. Chemosphere, **10**: 665-705.

Pilenkova II & Fatyanova AD (1980) [Determination of chlorendic anhydride in the air of working environments.] Zh Anal Khim, **35**: 2047-2049 (in Russian).

Schuphan I & Ballschmiter K (1972) Metabolism of polychlorinated norborenes by *Clostridium butyricum*. Nature (Lond), **237**(5350): 100-101.

Stowell JP & Jensen JN (1991) Dechlorination of chlorendic acid with ozone. Water Res, **25**(1): 83-90.

Talbot RC (1984) Using fiberglass-reinforced plastics. Chem Eng, **91**: 76-82.

Tryzna GE & Paa H (1975) Acute toxicity studies with crude Het anhydride. Northbrook, Illinois, Industrial Biotest Laboratories Inc. (Proprietary report No. 601.06668 to Hooker Chemical Corporation, Niagara Falls, USA).

Ulrich CE (1980) Chlorendic anhydride, subacute inhalation toxicity study in rats: 20 exposures in 28 days. Mattawan, Michigan, International Research and Development Corporation (Proprietary report No. 163.531, submitted to WHO by Velsicol Chemical Corporation, Chicago, USA).

Uno Y, Takasawa H, Miyagawa M, Inoue Y, Murata T, & Yoshikawa K (1994) An *in vivo-in vitro* replicative DNA synthesis (RDS) test using rat hepatocytes as an early prediction assay for non-genotoxic hepatocarcinogens screening of 22 known positives and 25 non carcinogens. Mutat Res, **320**(3): 189-205.

US EPA (1982) Final technical support document: Chlorendic acid (Contract No. 68-01-6530). Washington, DC, US Environmental Protection Agency, Office of Pesticides and Toxic Substances.

US ITC (1988) Synthetic organic chemicals - US production and sales, 1987. Washington, DC, US International Trade Commission, pp 3-9 (USITC Publication No. 2118).

US NTP (1987) Toxicology and carcinogenesis studies of chlorendic acid (CAS No. 115-28-6) in F344/N rats and $B6C3F_1$ mice (Feed studies). Research Triangle Park, North Carolina, US Department of Health and Human Services, National Toxicology Program (NTP Technical Report Series No. 304).

Velsicol Chemical Corporation (1982) Product information bulletin - Speciality chemicals: Velsicol chlorendic anhydride. Rosemont, Illinois, Velsicol Chemical Corporation.

Vilkas AG & Hutchinson CH (1977) Acute toxicity of chlorendic anhydride to the water flea, *Daphnia magna straus*. Tarrytown, New York, Union Carbide Environmental Services (Proprietary report No. 11506-03-06, submitted to WHO by Velsicol Chemical Corporation, Chicago, USA).

Witherup S, Stemmer KL, & Schlecht H (1965) The toxicity of chlorendic acid and chlorendic anhydride. Cincinnati, Ohio, University of Cincinnati, Department of Preventive Medicine and Industrial Health, Kethering Laboratory (Proprietary report submitted to WHO by Velsicol Chemical Corporation, Chicago, USA).

Ying WC, Bonk RR, & Sojka SA (1987) Treatment of a landfill leachate in powdered activated carbon enhanced sequencing batch bioreactors. Environ Prog, **6**(1): 1-8.

Yu CC & Atallah YH (1977a) Hydrolysis of chlorendic anhydride in aqueous solutions and stability on TLC plates. Chicago, Illinois, Velsicol Chemical Corporation (Proprietary report).

Yu CC & Atallah YH (1977b) Partition coefficients of MC-680, chlorendic anhydride and chlorendic acid. Chicago, Illinois, Velsicol Chemical Corporation (Proprietary report).

Yu CC & Atallah YH (1978) Photolysis of chlorendic anhydride and chlorendic acid. Chicago, Illinois, Velsicol Chemical Corporation (Proprietary report).

Zeiger E (1990) Mutagenicity of 42 chemicals in *Salmonella*. Environ Mol Mutagen, **16**(suppl 18): 32-54.

# RESUME ET EVALUATION; CONCLUSIONS ET RECOMMANDATIONS

## 1. Résumé et évaluation

### 1.1 Propriétés physiques et chimiques

L'acide chlorendique (qualité commerciale 99,5 %) et l'anhydride du même nom (qualité technique 97%) sont des substances cristallines voisines de couleur blanche. De par leur structure, ils sont apparentés aux insecticides diéniques chlorés. Par chauffage de l'acide chlorendique dans un système ouvert, on obtient l'anhydride correspondant. Inversement on peut rapidement revenir à l'acide par hydrolyse.Le point de fusion varie de 208°C pour l'acide à 235°C pour l'anhydride.

### 1.2 Production et usages

L'acide et l'anhydride chlorendiques sont surtout utilisés comme retardateurs de flamme dans les résines polyester et les plastifiants destinés aux installations électriques et aux peintures ainsi que dans les résines renforcées par des fibres de verre qui servent à la fabrication d'équipements pour l'industrie chimique. L'industrie textile les a aussi utilisés pour le finissage des tissus de laine et des tapis.

La production mondiale totale d'acide et d'anhydride chlorendiques tourne actuellement autour de 4 000 tonnes par an.

### 1.3 Transport, distribution et transformation dans l'environnement

Il peut y avoir libération d'acide chlorendique par suite d'une décomposition hydrolytique de polyesters ou de l'oxydation d'insecticides cyclodiéniques chlorés.

Le rayonnement ultraviolet décompose l'acide chlorendique avec une demi-vie de 16 jours lorsque ce dernier se trouve sous la forme d'une fine couche solide. La demi-vie tombe à 5 jours en solution aqueuse. Dans le sol,la demi-vie oscille entre 140 et 280 jours. Le composé est assez persistant dans le sol, mais on manque de données à ce sujet.

On ne sait rien non plus de son potentiel de bioaccumulation et de bioamplification. Par ailleurs, il n'y a aucune donnée sur la destinée finale des produits de la réaction de ce composé avec d'autres substances présentes dans les décharges et les incinérateurs.

L'exposition à l'anhydride chlorendique risque de conduire à une exposition à l'acide correspondant par suite de l'hydrolyse de l'anhydride.

## 1.4 Concentrations dans l'environnement et exposition humaine

On a trouvé de l'acide chlorendique à des concentrations atteignant 455 mg/litre dans l'eau de lessivage de certaines décharges.

## 1.5 Cinétique et métabolisme chez les animaux de laboratoire

Après administration d'acide chlorendique radiomarqué par voie orale à des rats, on a constaté que le composé se répartissait rapidement dans tout l'organisme et qu'il était métabolisé sans délai. Le marqueur radioactif a été excrété à plus de 90% dans les matières fécales en l'espace de 24 h, principalement sous forme de conjugué. Dans les urines, il n'a été excrété qu'à hauteur de 3 à 6%. C'est dans les tissus adipeux, le foie, les reins,le sang total et les poumons que l'on a mesuré les plus fortes teneurs en marqueur radioactif.

Des résultats analogues ont été obtenus lors d'une étude au cours de laquelle des rats ont reçu de l'anhydride chlorendique par gavage. Dans cette étude, on a obtenu une demi-vie de moins de 2 jours, sauf dans de cas du tissu adipeux où elle a atteint 22,5 jours.

On ne dispose d'aucune donnée sur la cinétique de ce composé après administration par la voie percutanée ou respiratoire.

## 1.6 Effets sur les mammifères de laboratoire et les systèmes d'épreuve *in vitro*

L'acide chorendique a une faible toxicité aiguë par voie orale; chez le rat, la $DL_{50}$ est égale à 1770 mg/kg de poids corporel. Dans le cas de l'anhydride, la $DL_{50}$ orale est de 2480 mg/kg de poids corporel.

La $DL_{50}$ aiguë par la voie percutanée est > 3000 mg/kg de poids corporel pour le lapin dans le cas de l'anhydride.

Pour l'acide chlorendique, la $CL_{50}$ à 4h en cas d'inhalation de poussières est > 0,79 mg/litre chez le rat.

Des tests sur des lapins ont montré que l'anhydride et l'acide chlorendiques ont un effet irritant cutané qui peut être sévère au niveau des yeux et des voies respiratoires.L'anhydride a un effet sensibilisateur sur la peau du cobaye, mais il est vrai que le même test effectué avec l'acide a donné un résultat négatif.

Lors d'une étude de 13 semaines sur des souris au cours de laquelle les animaux ont reçu de l'acide chlorendique mélangé à leur nourriture, on a obtenu la valeur de 2500 mg par kg de nourriture pour la dose sans effet observable, soit l'équivalent de 250 mg/kg de poids corporel. Chez des rats ayant fait l'objet d'une étude analogue, la valeur obtenue a été de 1250 mg par kg de nourriture, soit 62,5 mg/kg de poids corporel. Aux doses élevées, on a noté une diminution sensible de la croissance et l'examen microscopique a révélé des anomalies au niveau du foie.

Une étude d'alimentation de 90 jours, au cours de laquelle des rats ont reçu de l'anhydride chlorendique dans leur nourriture, a donné une valeur de 125 mg/kg de poids corporel par jour pour la dose sans effet observable. Dans une autre étude,qui a duré 3 semaines et a consisté à exposer les animaux par la voie cutanée, on a obtenu la valeur de 100 mg/kg de poids corporel par jour pour la dose en question (indépendamment de l'effet irritant pour la peau). En revanche, on n'a pas pu établir la dose sans effet observable lors d'une étude de 28 jours consistant à faire inhaler de la poussière à des rats.

Lors d'une étude de tératogénicité effectuée sur des rats, on a administré aux animaux de l'anhydride chlorendique par gavage à des doses allant jusqu'à 400 mg/kg de poids corporel du 6ème au 15ème jour de la gestation; le composé s'est révélé toxique pour les mères mais il n'y a pas eu d'effets tératogènes.

On a étudié le pouvoir mutagène de l'acide chlorendique sur cinq souches de *Salmonella typhimurium* en présence ou en l'absence d'un système métabolique exogène. Des résultats négatifs ont été obtenus avec des doses allant jusqu'à 7690 $\mu$g/boîte. Une épreuve de mutation sur lymphome murin s'est par contre révélée positive, en l'absence de système métabolique exogène. Les doses utilisées étaient respectivement égales à 1300 µg et 1700 µg/ml. La plus élevée des deux doses s'est révélée cytotoxique.

L'acide chlorendique a donné un résultat positif dans une épreuve de transformation sur des cellules BALB/c-3T3 en l'absence d'activation métabolique et un résultat négatif lors d'une épreuve sur des mâles de *Drosophila melanogaster*, pour la mise en évidence de mutations létales récessives liées au sexe. Le composé n'a pas provoqué d'augmentation de la synthèse réplicative de l'ADN après administration par voie orale ou sous-cutanée à des rats F-344 de doses respectivement égales à 450 ou 900 mg/kg de poids corporel.

Lors d'épreuves au cours desquelles de l'anhydride chlorendique a été testé sur cinq souches de *Salmonella typhimurium* et une souche de la levure *Saccharomyces cerevisiae* à des doses allant jusqu'à 7500 µg/boîte,on n'a pas constaté d'activité mutagène. Il n'a pas induit de mutations directes dans l'épreuve sur lymphome murin et a donné un résultat négatif lors d'une épreuve de transformation sur cellules BALB/3T3. L'anhydride chlorendique a également produit une synthèse non programmée de l'ADN en proportion non négligeable dans des cellules humaines WI-38. Lors d'une épreuve de létalité dominante, des souris ont été exposées à des doses uniques d'anhydride chlorendique allant jusqu'à 223 mg par kg de poids corporel, après une période de reproduction de 7 semaines. Le seul effet observé a été une réduction statistiquement significative de l'indice de fécondité, par rapport aux témoins, chez toutes les femelles accouplées avec des mâles traités, au cours de la 5ème semaine, ainsi que chez les femelles accouplées avec des mâles traités par une demi-dose, au cours de la 4ème semaine. Aucun autre effet n'a été observé, qu'il s'agisse du nombre des implantations, viables ou non viables, et des résorptions. Malgré les défauts de conception de cette étude, on en a conclu que les résultats de l'épreuve étaient négatifs.

On a étudié le pouvoir cancérogène de l'acide chlorendique sur des rats F-344/N à des doses respectivement égales à 620 et 1250 mg/kg de nourriture (soit l'équivalent de 31 et 62,5 mg/kg de poids corporel). Outre les anomalies non malignes observées dans un certain nombre d'organes-dégénérescence kystique et altérations focales au niveau cellulaire ou encore hyperplasie des canaux biliaires, on a noté une augmentation, significative à la dose maximale, de l'incidence des tumeurs chez les animaux traités: adénomes hépatocellulaires chez les mâles et adénomes ou carcinomes hépatocellulaires chez les femelles. De plus, il y avait chez les mâles une légère augmentation des cancers acineux du pancréas et des adénomes alvéolaires et bronchiolaires.

Chez des souris $B6C3F_1$ qui avaient reçu une alimentation contenant soit 620, soit 1250 mg d'acide chlorendique par kg de nourriture (c'est-à-dire l'équivalent de 62 ET 125 mg/kg de poids corporel, respectivement), on a constaté une plus grande incidence des nécroses et des altérations mitotiques au niveau du foie. Chez les mâles, il y a eu, aux deux doses, augmentation de l'incidence des adénomes et des carcinomes hépatocellulaires. Chez les femelles, c'est l'incidence des adénomes et des carcinomes alvéolaires et bronchiolaires qui était en augmentation.

On a cherché à élucider le mécanisme de la cancérogénèse en faisant appel à plusieurs modèles ou épreuves: l'épreuve d'initiation/promotion, le modèle avec hépatectomie partielle et le modèle néonatal. La conclusion a été que l'acide chlorendique se comporte comme un promoteur.

## 1.7 Effets chez l'homme

On ne dispose d'aucune donnée concernant d'éventuels effets sur l'homme.

## 1.8 Effets sur les autres êtres vivants au laboratoire et dans leur milieu naturel

L'acide chlorendique aurait des effets toxiques sur les algues à la dose de 250 mg/litre. Il s'agit notamment d'une inhibition de l'activité de la microfaune, d'une baisse de la production d'oxygène et d'une réduction de la respiration. On n'a en revanche signalé aucun effet toxique chez les algues exposées à 125 mg d'acide chlorendique par litre ou à de l'anhydride chlorendique. Les effets toxiques observés sont attribués à la modification du pH plutôt qu'à une toxicité directe de l'acide chlorendique. Aux faibles valeurs du pH, l'acide chorendique se trouve sous sa forme non ionisée, qui peut avoir un effet toxique direct.

Il n'y a qu'une seule espèce végétale terrestre pour laquelle on ait signalé des effets toxiques imputables à l'acide chlorendique. Après exposition à 0,1 mg/litre ou davantage, il y a eu inhibition de la croissance et de la germination, mais à la dose de 0,01 mg/litre, aucun effet n'a été signalé.

La $CL_{50}$ d'anhydride chlorendique pour *Daphnia magna* est de 110,7 mg/litre à 48h; elle est de 422,7 mg/litre à 96h pour la truite arc-en-ciel et *Lepomis macrochirus*.

Il est impossible d'évaluer les effets de l'anhydride chlorendique sur les êtres vivants dans leur milieu naturel sans connaître la concentration et la destinée de ce composé dans les différents compartiments de l'environnement.

## 2. Conclusions

La base de données relative à l'acide et à l'anhydride chlorendiques est loin d'être complète. Pour beaucoup d'études, le Groupe de travail ne disposait pas de rapports *in extenso*, mais seulement de résumés. En particulier, il n'y a aucune donnée sur la destinée ultime de ces composés, soit tels quels, soit après réaction avec d'autres substances et l'on ne sait pas non plus ce qu'il en est de leur potentiel de bioaccumulation et de bioamplification. Par ailleurs, on ne possède pas de données sur l'exposition de l'homme et des autres êtres vivants dans leur milieu naturel.

Les deux composés semblent n'avoir qu'une faible toxicité aiguë et subaiguë par voie orale, mais ils sont irritants pour l'oeil, la peau et les voies respiratoires. D'après les résultats des études de toxicité et de cancérogénicité à long terme, l'acide et l'anhydride chlorendiques provoquent la formation de tumeurs chez le rat et la souris, ce qui permet de les considérer comme potentiellement cancérogènes. On ne peut cependant pas évaluer le risque qu'ils représentent pour l'homme et son environnement, faute de données suffisantes.

La base de données actuelle n'est pas suffisamment étoffée pour que l'on puisse se prononcer en faveur d'une utilisation commerciale de l'acide et de l'anhydride chlorendiques.

# 3. Recommandations

## 3.1 Protection de la santé humaine et de l'environnement

a) L'exposition de la population générale à l'acide et à l'anhydride chlorendiques ainsi qu'à leurs dérivés doit être réduite au minimum.

b) Il ne faut utiliser l'acide et l'anhydride chlorendiques qu'en vase clos afin d'éviter une exposition aux vapeurs ou aux poussières. Les travailleurs qui produisent ou manipulent ces composés doivent être suffisamment familiarisés avec les mesures de sécurité. Des dispositifs techniques appropriés et des mesures relevant de l'hygiène et de la sécurité du travail doivent assurer leur protection.

c) L'évacuation de l'acide et de l'anhydride chlorendiques ou des déchets qui en contiennent doivent se faire selon des méthodes permettant d'éviter l'exposition de la population et de réduire au minimum la pollution de l'environnement.

## 3.2 Recherches futures

a) Il conviendrait de compléter la base de données par des renseignements suffisants sur:

   i) la destinée finale de ces composés, soit tels quels, soit après réaction;

   ii) leur potentiel de bioaccumulation et de bioamplification;

   iii) l'exposition humaine et la pollution de l'environnement;

b) Il faudrait étudier les produits de combustion de matériaux préparés ou traités avec de l'acide ou de l'anhydride chlorendiques et, en particulier, leur toxicité par la voie respiratoire ainsi que le risque de pollution de l'environnement par ces produits.

c) Il conviendrait de mesurer la concentration de l'acide et de l'anhydride chlorendiques dans les différents compartiments de l'environnement ou de s'efforcer tout au moins de les calculer à l'avance à l'aide des modèles existants.

d) Il faudrait obtenir des données sur la présence éventuelle d'aberrations chromosomiques en procédant à une analyse de la métaphase. Il faudra également obtenir des données de mutagénicité *in vivo* avant de pouvoir procéder à une analyse complète du risque.

e) Une étude de reproduction portant sur trois générations serait à effectuer.

f) Il faudrait élucider le mécanisme qui est à la base de la cancérogénicité de ces deux substances.

# RESUMEN Y EVALUACION; CONCLUSIONES Y RECOMENDACIONES

## 1. Resumen y evaluación

### 1.1 Propiedades físicas y químicas

El ácido cloréndico (calidad comercial, 99,5%) y el anhídrido cloréndico (calidad técnica, 97%) son productos cristalinos blancos estrechamente relacionados entre sí. Estructuralmente son muy parecidos a los insecticidas clorados derivados del ciclodieno. Calentado en un sistema abierto, el ácido cloréndico pierde agua y forma anhídrido cloréndico. Este puede hidrolizarse rápidamente y transformarse en ácido cloréndico. Los puntos de fusión van de 208 °C (para el ácido) a 235 °C (para el anhídrido).

### 1.2 Producción y uso

El ácido cloréndico y el anhídrido cloréndico se utilizan principalmente como reactivos pirorretardantes en resinas poliestéricas y agentes ablandantes para sistemas eléctricos y pinturas, y en resinas reforzadas con fibra de vidrio para equipo de procesos químicos. En la industria textil se utilizaron en el pasado en el acabado de alfombras y tejidos de lana.

La producción mundial total de ácido cloréndico y anhídrido cloréndico se sitúa actualmente en torno a las 4000 toneladas por año.

### 1.3 Transporte, distribución y transformación en el medio ambiente

El ácido cloréndico puede liberarse por degradación hidrolítica de poliésteres y como producto de la oxidación de los insecticidas clorados derivados del ciclodieno.

La radiación ultravioleta degrada el ácido cloréndico, cuya semivida es de 16 días en una capa sólida fina y de 5 días en solución acuosa. En el suelo la semivida oscila entre 140 y 280 días. El ácido cloréndico es bastante persistente en el suelo, aunque los datos disponibles a este respecto son insuficientes.

No se dispone de datos sobre el potencial de bioacumulación y bioamplificación. Asimismo, se carece de datos sobre el destino final de los productos de reacción entre otros en la eliminación de desechos y la incineración.

La exposición al anhídrido cloréndico entraña probablemente también la exposición al ácido cloréndico, debido a la hidrólisis del primero.

## 1.4 Niveles medioambientales y exposición humana

En el lixiviado de vertederos se ha encontrado ácido cloréndico a concentraciones de hasta 455 mg/litro.

## 1.5 Cinética y metabolismo en animales de laboratorio

Tras la administración a ratas de ácido cloréndico radiomarcado por vía oral e intravenosa, la sustancia se distribuyó velozmente por todo el organismo y se metabolizó con rapidez. Más del 90% se excretó en las primeras 24 horas en las heces, principalmente en forma conjugada. Sólo entre el 3% y el 6% se excretó en la orina. Las concentraciones más altas de la sustancia radiomarcada se hallaron en el tejido adiposo, el hígado, los riñones, la sangre entera y los pulmones.

En un estudio de alimentación con sonda en ratas se obtuvieron resultados parecidos con el anhídrido cloréndico. La semivida del compuesto marcado en este estudio fue inferior a dos días, salvo en la grasa, donde ascendió a 22,5 días.

No se dispone de datos sobre la cinética tras la exposición cutánea o respiratoria.

## 1.6 Efectos en mamíferos de laboratorio y en sistemas de pruebas *in vitro*

La toxicidad oral aguda del ácido cloréndico es baja; la $DL_{50}$ en la rata es de 1770 mg/kg de peso corporal. Para el anhídrido cloréndico, la $DL_{50}$ por vía oral en las ratas es de 2480 mg/kg de peso corporal.

La $DL_{50}$ aguda por vía cutánea del anhídrido cloréndico crudo en el conejo es de > 3000 mg/kg de peso corporal.

En las ratas, la $CL_{50}$ del ácido cloréndico por inhalación de polvo durante cuatro horas es de > 0,79 mg/litro.

El ácido y el anhídrido cloréndico provocan irritación cutánea y grave irritación de los ojos y de las vías respiratorias en el conejo. El anhídrido cloréndico produce sensibilización de la piel en el cobayo, mientras que una prueba con ácido cloréndico fue negativa.

En un estudio de alimentación de ratones de 13 semanas de duración en el que se utilizó ácido cloréndico se halló un nivel sin efectos observados (NOEL) de 2500 mg/kg de alimento (equivalente a 250 mg/kg de peso corporal); en un estudio de alimentación análogo realizado con ratas el NOEL fue de 1250 mg/kg de alimento (equivalente a 62,5 mg/kg de peso corporal). A dosis más altas la disminución del crecimiento fue significativa y se observaron cambios microscópicos en el hígado.

En un estudio de alimentación de 90 días en ratas con anhídrido cloréndico, el NOEL fue de 125 mg/kg de peso corporal por día, y en otro estudio de tres semanas con administración por vía cutánea el NOEL ascendió a 100 mg/kg de peso corporal por día (aparte de la irritación de la piel). En un estudio de inhalación de polvo en ratas durante 28 días no se pudo establecer ningún NOEL.

Un estudio de teratogenicidad en la rata con anhídrido cloréndico, administrado mediante alimentación con sonda a dosis de hasta 400 mg/kg de peso corporal en los días 6 a 15 de la gestación, reveló toxicidad materna pero no efectos teratógenos.

Se determinó el potencial mutagénico del ácido cloréndico en cinco cepas de *Salmonella typhimurium* en presencia y ausencia de un sistema de metabolismo exógeno. Con dosis de hasta 7690 μg/placa se obtuvieron resultados negativos. Una prueba de mutación en células de linfoma de ratón en ausencia de un sistema de metabolismo exógeno fue positiva. Las dosis utilizadas fueron de 1300 a 1700 μg/ml. La dosis más alta resultó ser citotóxica.

El ácido cloréndico dio resultados positivos en una prueba de transformación con células BALB/c-3T3 sin activación metabólica, y negativo en una prueba de mutaciones letales recesivas ligadas al sexo en machos de *Drosophila melanogaster*. El ácido cloréndico no estimuló la síntesis replicativa del ADN tras la aplicación oral o subcutánea de 450 ó 900 mg/kg de peso corporal a ratas F-344.

En pruebas realizadas en cinco cepas de *Salmonella typhimurium* y una de la levadura *Saccharomyces cerevisiae* con dosis de hasta 7500 µg/placa, el anhídrido cloréndico no resultó potencialmente mutagénico. El compuesto no indujo mutaciones directas en un ensayo con células de linfoma de ratón, y no tuvo efectos en una prueba de transformación con células BALB/3T3. En células humanas WI-38 produjo un grado significativo de síntesis no programada de ADN. En una prueba de dominancia letal en ratones, se les expuso a dosis únicas de hasta 223 mg de anhídrido cloréndico/kg de peso corporal, seguidas de un periodo de reproducción de siete semanas. Sólo se observó una disminución estadísticamente significativa del índice de fecundidad, respecto de los testigos, en todas las hembras que se aparearon con machos tratados durante la semana 5, y en las que se aparearon con machos tratados una dosis media durante la semana 4. No se detectaron efectos en el número de implantaciones, resorciones o huevos muertos. Se concluyó que esta prueba daba resultados negativos, si bien el diseño del estudio era inadecuado.

El potencial carcinogénico del ácido cloréndico se determinó en ratas F-344/N a dosis de 620 y 1250 mg/kg de alimento (equivalentes a 31 y 62,5 mg/kg de peso corporal). Además de algunas modificaciones no neoplásicas significativas en varios órganos, como la degeneración quística y alteraciones celulares focales, y de una hiperplasia del conducto biliar en el hígado, se observaron aumentos de la incidencia de adenomas hepatocelulares en los machos tratados, y de adenomas y carcinomas hepatocelulares, significativos a las dosis más altas, en las hembras. Además, en los machos se detectaron ligeros aumentos de los adenomas de células acinosas del páncreas y de los adenomas de alveolos/bronquiolos en el pulmón.

En ratones $B6C3F_1$ alimentados con dietas que contenían 620 y 1250 mg de ácido cloréndico por kg de alimento (equivalentes a 62 y 125 mg/kg de peso corporal) se observó una mayor incidencia de necrosis y alteraciones mitóticas en el hígado. Con ambas dosis se observó un aumento de los adenomas y carcinomas hepatocelulares en los machos. En las hembras se halló una mayor incidencia de adenomas o carcinomas de los alveolos/bronquiolos.

Se realizaron estudios para investigar los mecanismos de carcinogénesis mediante una valoración del potencial iniciador/facilitador, el modelo de hepatectomía parcial y el modelo neonatal. Las pruebas indicaron que el ácido cloréndico tiene actividad facilitadora.

### 1.7 Efectos en el ser humano

No se dispone de datos sobre los efectos en el ser humano.

### 1.8 Efectos en otros organismos en el laboratorio y en el medio ambiente

Se ha notificado que a 250 mg/litro el ácido cloréndico tiene efectos tóxicos sobre las algas. Entre los efectos señalados figuran la inhibición de la actividad de la microfauna, la menor producción de oxígeno y la merma de la respiración. Para las algas expuestas a 125 mg de ácido cloréndico/litro y para las expuestas al anhídrido cloréndico no se han notificado efectos tóxicos. Los efectos tóxicos observados en las algas se han atribuido a la modificación del pH más que a la toxicidad directa del ácido cloréndico. A valores de pH bajos el ácido cloréndico está en forma no ionizada, forma que puede tener efectos tóxicos directos.

Se han señalado efectos del ácido cloréndico sólo sobre una especie de planta terrestre. Tras la exposición a 0,1 mg/litro o más, se observó una inhibición tanto del crecimiento como de la germinación de las semillas, mientras que la exposición a 0,01 mg/litro no produjo ningún efecto.

La $CL_{50}$ a las 48 horas para *Daphnia magna* fue de 110,7 mg de anhídrido cloréndico/litro, y la $CL_{50}$ a las 96 horas para la trucha arco iris y para *Lepomis machrochirus* fue de 422,7 mg/litro.

Los efectos potenciales del anhídrido cloréndico sobre los organismos en el medio ambiente no pueden evaluarse en ausencia de datos sobre las concentraciones y los procesos que determinan el destino de este compuesto en los compartimientos ambientales.

## 2. Conclusiones

La base de datos sobre el ácido cloréndico y el anhídrido cloréndico dista mucho de ser completa. Para varios estudios el Grupo de Trabajo no dispuso de la versión integral de los informes (sino sólo de resúmenes). En particular, no hay datos sobre el destino final de las sustancias en sí ni de los productos a que dan lugar, ni tampoco sobre el potencial de bioacumulación y bioamplificación. Además, se carece de datos sobre la exposición de seres humanos y de organismos en el medio ambiente.

Las dos sustancias parecen tener una toxicidad aguda y subaguda baja por vía oral, pero son irritantes de la piel, los ojos y las vías respiratorias. Los resultados de estudios de toxicidad/carcinogenicidad de larga duración con administración de ácido cloréndico a ratas y ratones llevan a concluir que el ácido cloréndico provoca tumores en ambos, por lo que se considera que tiene un potencial carcinógeno. Sin embargo, vista la falta de datos, no es posible hacer una evaluación cabal de la peligrosidad para el ser humano y el medio ambiente.

La base de datos actual es insuficiente para respaldar el uso comercial del ácido cloréndico y del anhídrido cloréndico.

## 3. Recomendaciones

### 3.1 Protección de la salud humana y del medio ambiente

a) La exposición de la población general al ácido cloréndico, al anhídrido cloréndico y a los productos de ellos derivados debe reducirse al mínimo.

b) El ácido cloréndico y el anhídrido cloréndico deben utilizarse sólo en sistemas cerrados para evitar la exposición al vapor y al polvo. Los trabajadores que producen y manipulan estas sustancias deben estar debidamente adiestrados en los procedimientos de seguridad, y se les debe proteger de la exposición con los adecuados medios técnicos de control y con medidas apropiadas de higiene industrial.

c) La eliminación del ácido y el anhídrido cloréndico y de sus productos residuales debe efectuarse por métodos que aseguren que la población general no quede expuesta y que la exposición del medio ambiente sea mínima.

### 3.2 Nuevas investigaciones

a) La base de datos debería completarse con información adecuada sobre:

i) el destino final de estas sustancias como tales y de sus productos de reacción;

ii) el potencial de bioacumulación y de bioamplificación;

iii) la exposición humana y ambiental.

b) Deberían realizarse estudios sobre los productos de combustión de los materiales preparados o tratados con el ácido o el anhídrido, su toxicidad por inhalación, y su potencial de contaminación del medio ambiente.

c) Las concentraciones de ácido y anhídrido cloréndico en los compartimientos ambientales deberían medirse, o por lo menos predecirse con ayuda de los modelos disponibles.

d) Deberían obtenerse datos sobre la presencia de aberraciones cromosómicas mediante un estudio de análisis de la metafase. Se necesitarán algunos datos sobre la mutagenicidad *in vivo* antes de poder efectuar una evaluación cabal de la peligrosidad.

e) Debería realizarse un estudio de reproducción de tres generaciones.

f) Debería clarificarse el mecanismo carcinogénico de ambas sustancias.

Fenvalerate (No. 95, 1990)
Fluorine and fluorides (No. 36, 1984)
Food additives and contaminants in food, principles for the safety assessment of (No. 70, 1987)
Formaldehyde (No. 89, 1989)
Genetic effects in human populations, guidelines for the study of (No. 46, 1985)
Glyphosate (No. 159, 1994)
Guidance values for human exposure limits (No. 170, 1994)
Heptachlor (No. 38, 1984)
Hexachlorobutadiene (No. 156, 1994)
Alpha- and beta-hexachlorocyclohexanes (No. 123, 1992)
Hexachlorocyclopentadiene (No. 120, 1991)
*n*-Hexane (No. 122, 1991)
Hydrazine (No. 68, 1987)
Hydrogen sulfide (No. 19, 1981)
Hydroquinone (No. 157, 1994)
Infancy and early childhood, principles for evaluating health risks from chemicals during (No. 59, 1986)
Inorganic lead (No. 165, 1995)
Isobenzan (No. 129, 1991)
Isophorone (No. 174, 1995)
Kelevan (No. 66, 1986)
Lasers and optical radiation (No. 23, 1982)
Lead (No. 3, 1977)[a]
Lead - environmental aspects (No. 85, 1989)
Lindane (No. 124, 1991)
Linear alkylbenzene sulfonates and related compounds (No. 169, 1996)
Magnetic fields (No. 69, 1987)
Man-made mineral fibres (No. 77, 1988)
Manganese (No. 17, 1981)
Mercury (No. 1, 1976)[a]
Mercury - environmental aspects (No. 86, 1989)
Mercury, inorganic (No. 118, 1991)
Methomyl (No. 178, 1996)
2-Methoxyethanol, 2-ethoxyethanol, and their acetates (No. 115, 1990)
Methyl bromide (No. 166, 1995)
Methylene chloride
(No. 32, 1984, 1st edition)
(No. 164, 1996, 2nd edition)
Methyl ethyl ketone (No. 143, 1992)
Methyl isobutyl ketone (No. 117, 1990)
Methylmercury (No. 101, 1990)
Methyl parathion (No. 145, 1992)
Mirex (No. 44, 1984)
Morpholine (No. 179, 1996)
Mutagenic and carcinogenic chemicals, guide to short-term tests for detecting (No. 51, 1985)
Mycotoxins (No. 11, 1979)
Mycotoxins, selected: ochratoxins, trichothecenes, ergot (No. 105, 1990)
Nephrotoxicity associated with exposure to chemicals, principles and methods for the assessment of (No. 119, 1991)
Neurotoxicity associated with exposure to chemicals, principles and methods for the assessment of (No. 60, 1986)
Nickel (No. 108, 1991)
Nitrates, nitrites, and *N*-nitroso compounds (No. 5, 1978)[a]
Nitrogen, oxides of (No. 4, 1977)[a]
2-Nitropropane (No. 138, 1992)
Noise (No. 12, 1980)[a]
Organophosphorus insecticides: a general introduction (No. 63, 1986)
Paraquat and diquat (No. 39, 1984)
Pentachlorophenol (No. 71, 1987)
Permethrin (No. 94, 1990)
Pesticide residues in food, principles for the toxicological assessment of (No. 104, 1990)
Petroleum products, selected (No. 20, 1982)
Phenol (No. 161, 1994)
d-Phenothrin (No. 96, 1990)
Phosphine and selected metal phosphides (No. 73, 1988)
Photochemical oxidants (No. 7, 1978)
Platinum (No. 125, 1991)
Polybrominated biphenyls (No. 152, 1994)
Polychlorinated biphenyls and terphenyls
(No. 2, 1976, 1st edition)[a]
(No. 140, 1992, 2nd edition)
Polychlorinated dibenzo-*p*-dioxins and dibenzofurans (No. 88, 1989)
Progeny, principles for evaluating health risks associated with exposure to chemicals during pregnancy (No. 30, 1984)
1-Propanol (No. 102, 1990)
2-Propanol (No. 103, 1990)
Propachlor (No. 147, 1993)
Propylene oxide (No. 56, 1985)
Pyrrolizidine alkaloids (No. 80, 1988)

---

[a] Out of print